Design for Six Sigma
in Product and
Service Development

Applications and Case Studies

Design for Six Sigma in Product and Service Development

Applications and Case Studies

Edited by
Elizabeth A. Cudney and Sandra L. Furterer

CRC Press
Taylor & Francis Group
Boca Raton London New York

CRC Press is an imprint of the
Taylor & Francis Group, an **informa** business

CRC Press
Taylor & Francis Group
6000 Broken Sound Parkway NW, Suite 300
Boca Raton, FL 33487-2742

First issued in paperback 2019

© 2012 by Taylor & Francis Group, LLC
CRC Press is an imprint of Taylor & Francis Group, an Informa business

No claim to original U.S. Government works

ISBN-13: 978-1-4398-6060-1 (hbk)
ISBN-13: 978-0-367-38126-4 (pbk)

Visit the Taylor & Francis Web site at
http://www.taylorandfrancis.com

and the CRC Press Web site at
http://www.crcpress.com

I would like to dedicate this book to my mother, who has always given to me. Whether it was support, time, or of herself, she has been a constant role model in living a gratifying life for others. I would also like to dedicate this book to my late father, who set an example for me on how to find balance personally and professionally.

Finally, I would like to dedicate this book to my children, Caroline and Joshua, and my husband Brian. My children never cease to amaze me and give me joy, support, and inspiration. I am truly lucky to have such a supportive and loving husband and family.

Beth Cudney

I dedicate this book to my loving family, my husband Dan, and my children Kelly, Erik, and Zachary, who provide constant support and the true purpose in my life. My children provide the incentive for me to strive for excellence to be a guiding example for their lives. A special thanks to my son Zachary for staying up late with me as I wrote and edited the book.

I also dedicate this book to my parents, Joan and the late Mel Brumback, who instilled in me the value of education and a lifelong love of books and learning, and also to my siblings, Dan, Neil, Tim, and Kathy, who encouraged the competitive spirit and being our best in all we do.

Sandy Furterer

Contents

Section I   Product Design for Six Sigma Projects

Section II Service Process Design for Six Sigma Projects

Preface

This book grew out of the need for our students to better understand how to apply and integrate the Design for Six Sigma methodology and its associated tools and techniques. Real-world examples and hands-on experience are invaluable resources when instructing the use of methods and tools in training or in a classroom. The instructors may not have access to these resources, thus they can teach only theory and basic examples. Another solution is the use of case studies. Case studies can help enhance the learning experience by allowing the learner a role in a real scenario. The story of the case study adds life to a seemingly lifeless group of tools. With this understanding, Design for Six Sigma methods taught with the aid of case studies may help some novices and even experienced participants better assimilate the tools, because they are presented as a whole.

The primary objective of this book is to provide real-world Design for Six Sigma projects performed for a wide variety of product and process/ service designs following the Identify-Define-Design-Optimize-Validate (IDDOV) process. The purpose is to facilitate Design for Six Sigma instruction by providing interactive case studies that will enable learners to apply the IDDOV phases.

This book would not be possible without the enthusiasm, dedication, and energy that our Design for Six Sigma students and team members exhibit. Our goal as authors and editors is to provide the learner with an understanding of how others applied Design for Six Sigma and a guide for how they might solve their organization's problems by applying Design for Six Sigma.

Acknowledgments

Our thanks and appreciation go to all of the Design for Six Sigma team members, project champions, and mentors who worked so diligently and courageously on these projects.

Special thanks go to several people for their contributions to the development and production of the book at CRC Press/Taylor and Francis, including Cindy Renee Carelli (Senior Editor), Amy Blalock (Project Coordinator), and Michael Davidson (Project Editor).

About the Editors

Beth Cudney, Ph.D., is an assistant professor in the Engineering Management and Systems Engineering Department at Missouri University of Science and Technology, Rolla, Missouri. Cudney received her B.S. in Industrial Engineering from North Carolina State University. She received her Master of Engineering in Mechanical Engineering with a Manufacturing Specialization and Master of Business Administration from the University of Hartford, and her doctorate in Engineering Management from the University of Missouri–Rolla. Her doctoral research focused on pattern recognition and developed a methodology for prediction in multivariate analysis. Cudney's research was recognized with the 2007 American Society for Engineering Management (ASEM) Outstanding Dissertation Award.

Prior to returning to school for her doctorate, she worked in the automotive industry in various roles including Six Sigma Black Belt, Quality/Process Engineer, Quality Auditor, Senior Manufacturing Engineer, and Manufacturing Manager. Cudney is an ASQ Certified Six Sigma Black Belt, Certified Quality Engineer, Certified Manager of Quality/Operational Excellence, Certified Quality Inspector, Certified Quality Improvement Associate, Certified Quality Technician, and Certified Quality Process Analyst. She is a past president of the Rotary Club of Rolla, Missouri. She is a member of the Japan Quality Engineering Society (QES), American Society for Engineering Education (ASEE), American Society of Engineering Management (ASEM), American Society of Mechanical Engineers (ASME), American Society for Quality (ASQ), Institute of Industrial Engineers (IIE), and Society of Automotive Engineers (SAE).

In 2010, Cudney was inducted as an associate member into the American Society for Quality's International Academy for Quality. In addition, she received the 2007 American Society for Quality (ASQ) Armand V. Feigenbaum Medal. This international award is given annually to one individual "who has displayed outstanding characteristics of leadership, professionalism, and potential in the field of quality and also whose work has been or, will become of distinct benefit to mankind." In addition, she received the 2006 Society of Manufacturing Engineers (SME) Outstanding Young Manufacturing Engineer Award. This international award is given annually to engineers "who have made exceptional contributions and accomplishments in the manufacturing industry."

Cudney has published over 50 conference papers and more than 25 journal papers. Her first book entitled, *Using Hoshin Kanri to Improve the Value Stream*, was released in March 2009 through Productivity Press, a division of Taylor and Francis. Her second book entitled, *Implementing Lean Six Sigma throughout the Supply Chain: The Comprehensive and Transparent Case Study*, was released in November 2010 through Productivity Press, a division of Taylor and Francis.

Sandy Furterer, Ph.D., MBA, CSSBB, Six Sigma Master Black Belt, CQE, is developing and deploying the Enterprise Performance Excellence program at an acute-care hospital. Prior to healthcare, she was a business architect in the Information Systems Division, Business and Process Architecture Team, with a major retailer. Furterer lead the Business and Process Architecture team in the Retail Systems application development area implementing best practices to achieve operational excellence in information systems application development processes, and helping to enable the business through application of business architecture and process improvement. She also was a Master Black Belt in the retailer's Global Continuous Improvement practice.

Furterer received her Ph.D. in Industrial Engineering with a specialization in Quality Engineering from the University of Central Florida, in Orlando, in 2004. She developed a state-of-the-art framework and roadmap for integrating Lean and Six Sigma for service industries and implemented the framework in a local government's financial services department. She received an MBA from Xavier University in Cincinnati, Ohio, and a Bachelor and Master of Science in Industrial and Systems Engineering from the Ohio State University in Columbus, Ohio.

Furterer was an assistant professor in the Industrial Distribution and Logistics program at East Carolina University from 2006 to 2007. She was a visiting assistant professor and assistant department chair in the Industrial Engineering and Management Systems Department at the University of Central Florida from 2004 to 2006.

She is currently a part-time faculty member in the Masters of Science in Quality Assurance program at Southern Polytechnic State University. She is also a graduate scholar at the University of Central Florida.

Furterer has over 18 years experience in business process and quality improvements. She is an ASQ Certified Six Sigma Black Belt, Certified Quality Engineer, and a certified Harrington Institute Master Black Belt. Furterer is an experienced consultant who has facilitated and implemented quality, statistics, and process improvement projects, using Six Sigma and Lean principles and tools. She has helped Fortune 100 companies, local governments, and nonprofit organizations streamline their processes and implement information systems.

She has published and/or presented over 20 conference papers/proceedings in the areas of Lean Six Sigma, quality, operational excellence, business architecture, and engineering education. Furterer is the editor of *Lean Six Sigma in Service: Applications and Case Studies* (CRC Press, 2009), *Lean Six Sigma in Healthcare: Applications and Case Studies* (CRC Press), and *Systems Engineering Focus to Business Architecture: Models, Methods, and Applications* (CRC Press).

Furterer is a senior member in the Institute of Industrial Engineers and a senior member in the American Society for Quality. She lives in Coral Springs, Florida, with her husband and three children.

List of Design for Six Sigma Project Team Members and Contributing Authors

Anas AlGhamri

Roger Ates

James Baker

Prajakta Bhagwat

Jason Castro

Austin Das

Priya Dhuwalia

David Ding

Benjamin Dowell

Thomas Fitzpatrick

Josef Garcia

Robert Jackson

Sumant Joshi

Neil Kester

Jamison Kovach

Colby Krug

Bruce Lane

Patricia Long

Patrick McCarthy

Soundararajan Chandra Moleeswaran

David Moore

Kayla Najjar

Vishwanath Narayanan

Nick Paul

Swathi Priya Pedavalli

Sagnik Saha

Adam Samiof

Lisette Sandoval

Çiğdem Sheffield

Amogh Shenoy

Antonio Ward

1

Instructional Strategies for Using This Book

Elizabeth Cudney and Sandra Furterer

CONTENTS

The purpose of this book is to provide a guide for learners and appliers of Design for Six Sigma methodologies and tools. Design for Six Sigma (DFSS) is a roadmap for the development of robust products and services. DFSS is a data-driven quality strategy for designing products and services. It is an integral part of the Six Sigma quality initiative. The goal of DFSS is to avoid manufacturing/service process problems using systems engineering techniques. DFSS consists of five interconnected phases: Identify, Define, Design, Optimize, Validate (IDDOV). As such, it is a closed-loop system that starts and ends with the customer.

This book provides a detailed description of how to apply Design for Six Sigma in product and service development. The book discusses the Design for Six Sigma roadmap that links several methodologies including organizational leadership, product development, system integration, critical parameter management, voice of the customer, quality function deployment, and concept generation, among others.

The book provides several real-world case studies and applications of Design for Six Sigma which have shown significant improvement in meeting customer requirements following the IDDOV methodology. The case studies include a variety of new products and services. These cases can be used by both industry professionals and academics to learn how to apply Design for Six Sigma. Examples of tools include Quality Function Deployment, Voice of the Customer, Pugh Concept Selection, Ideal Function, Failure Modes and Effects Analysis, Reliability, Measurement Systems Analysis, Regression Analysis, and Capability Studies, among others. The case studies will benefit the reader by showing the tools and how to integrate them for robust product and service design. The case studies will provide a detailed, step-by-step approach to Design for Six Sigma.

The book is designed to engage the reader by enabling hands-on experience with real Design for Six Sigma project cases in a safe environment, where experienced Black Belts and Master Black Belts can help mentor the students in Design for Six Sigma. Case studies are designed to enable students to work through the exercises and to provide sufficient background information so that they can apply the tools as if they collected the data themselves. The case discussions provide questions to allow students to compare their solutions with actual results and decisions realized by similar students who are learning and applying Design for Six Sigma. This will help prepare them to touch actual data and make decisions when they embark on real-world projects.

1.1 Business Process and Lean Six Sigma Project Backgrounds

The Design for Six Sigma case studies consist of product design and service-oriented processes in various environments. An overview of each project is provided for the students so that they understand the background of the project, as well as sufficient information regarding the products and services that need to be improved so that they can develop a project charter and scope the project. Data that were actually collected in the Design for Six Sigma projects are provided for application of Design for Six Sigma tools and appropriate statistical analysis. Case exercises are provided so that the students can solve the Design for Six Sigma projects for each phase of the Identify-Define-Design-Optimize-Validate (IDDOV) or Define-Measure-Analyze-Design-Verify (DMADV) problem-solving methodology. Each phase provides the solutions the students actually developed that can be used as a guide to solve the next phase of the project.

1.2 Design for Six Sigma Case Study Goals

To successfully complete the Design for Six Sigma case studies, participants must apply appropriate problem-solving methods and tools from the Design for Six Sigma toolkit to understand the problem, identify key customers and stakeholders, understand critical to satisfaction (CTS) characteristics, develop potential process and product concepts, select superior concepts, optimize the process or product, and develop a plan to mitigate risk for the new process or product.

1.3 Design for Six Sigma Tools

During the case study, the class will use Design for Six Sigma IDDOV or DMADV and project planning and statistical software that are most commonly used in real-world projects.

1.4 Learning Design

Each exercise in the case study is designed so that the teams of students experience the factors listed below:

- Team interaction, definition of team ground rules, brainstorming, and consensus building, as well as the stages of team growth
- Choice of how to apply Design for Six Sigma tools and problem-solving methods
- Support of their decisions and application of the tools with data
- Review of information for relevant and irrelevant information and data, and reframing into what is important to solve the problem
- With each exercise, development of students' understanding and application of specific tools and problem-solving methods
- Development of written and oral communication through customer interaction and written reports and presentations, as well as the ability to present technical information
- Application of project management tools to manage activities and complete tasks in a timely manner
- Experience in solving an unstructured problem in a safe learning environment where mentoring is available

The instructor's role is to facilitate the learning process. It is critical for the instructor to act as a coach or mentor to the student teams. It can also be helpful to have Six Sigma Black Belts or Master Black Belts experienced in applying Design for Six Sigma tools and methods assigned to each student team to mentor them in the application of Design for Six Sigma projects.

Most Design for Six Sigma programs work on projects in teams. Therefore, the instructor can organize the students into teams of four to six students. There is a great deal of value in having students work together as a team to work on a DFSS project. They can learn how to work more effectively as a team, and team members can transfer learning across the team members because students grasp the difficult concepts of Design for Six Sigma at different paces. An effective way to organize the teams is to determine the students' experience and balance the team with a group ranging from no experience to extensive experience.

1.5 Required Knowledge Levels by Design for Six Sigma Projects

The Design for Six Sigma projects included in this book stem from different knowledge levels and depth of understanding for applying DFSS tools and techniques. There are three different student levels defined as follows:

- Beginner—Undergraduate student (usually senior) student with no exposure to Design for Six Sigma and little statistical background
- Intermediate—Master's student with some exposure (theoretical knowledge) to Design for Six Sigma tools and some statistical background
- Advanced—Master's or Ph.D. graduate students with theoretical learning of Design for Six Sigma tools and some statistical background, as well as having worked on a Design for Six Sigma or Six Sigma project

This book is divided into two parts to illustrate case studies in product and service design. Part 1 provides product design cases (Chapters 3 through 11). Part 2 provides service/process design case studies (Chapters 12 through 14).

2

Design for Six Sigma Identify-Define-Design-Optimize-Validate (IDDOV) Roadmap Overview

Sandra Furterer and Elizabeth Cudney

CONTENTS

2.1 Design for Six Sigma Overview

Design for Six Sigma (DFSS) is a methodology that can be used to systematically design new products, services, or processes. It embeds the underlying management philosophies and principles and the stretch goal of the Six Sigma DMAIC (Define-Measure-Analyze-Improve-Control) methodology. DFSS focuses on designing a product, service, or process right the first time so less time needs to be spent downstream in improving the product, service, or process. We will discuss Design for Six Sigma and the IDDOV (Identify-Define-Design-Optimize-Validate) methodology as they relate to product design and service-oriented and transaction-based processes.

In Subir Chowdhury's book, *Design for Six Sigma*, he believes that Six Sigma can only take an organization so far, and that organizations must focus on designing good products and processes, so there is less need to improve them, which can prevent errors from occurring (Chowdhury, 2002). From Deming's quality and profitability cycle, improved quality of design can lead to higher perceived value by the customer, which can contribute to increased market share, margins, revenue, and profitability (Deming, 2000). Most organizations can reach a level of 4.5 sigma through applying Six Sigma projects for process improvement. It is approximately at that point when organizations hit a brick wall with improvements and must turn to Design for Six

Sigma to further improve products and services through design or redesign of those products or services.

Unlike Lean Six Sigma, which typically uses the DMAIC problem-solving methodology, Design for Six Sigma literature discusses applying many different methodologies to design the new products or services, such as DMADV (Design-Measure-Analyze-Design-Validate), IDDOV (Identify-Define-Design-Optimize-Validate), IDOV (Identify-Design-Optimize-Validate), DMADOV (Design-Measure-Analyze-Design-Optimize-Verify). The authors adapted the IDDOV methodology discussed by Chowdhury and developed the roadmap for applying Design for Six Sigma using IDDOV to product- and service-oriented process design.

The IDDOV process could be used when creating a brand new process that has never been done before in your organization or to make a major redesign of an existing process. This existing process may be too broken to provide guidance for the redesign. The same methodology can be applied to design a product as well. The IDDOV process provides a comprehensive set of tools and techniques that can be used based on the type of project (i.e., product design, redesign).

The benefits of applying Design for Six Sigma and IDDOV compared to Six Sigma and DMAIC are that you are not constrained by an existing process, and you do not need to collect large amounts of voice of process (VOP) data, or spend time baselining a nonexistent or seriously broken process.

Figure 2.1 illustrates the steps that are part of each phase of the IDDOV methodology. Figure 2.2 maps the tools typically used in each phase of the IDDOV methodology.

The following sections provide a roadmap of how to apply IDDOV and the main tools that could be applied when designing a service process.

Identify	Define	Design	Optimize	Validate
1. Develop project charter	4. Collect voice of the customer (VOC)	7. Identify process elements	10. Implement pilot process	13. Validate process
2. Perform stakeholder analysis	5. Identify critical to satisfaction (CTS) measures and targets	8. Design process	11. Assess process capabilities	14. Assess performance, failure modes, and risks
3. Develop project plan	6. Translate VOC into technical requirements	9. Identify potential risks and inefficiencies	12. Optimize design	15. Iterate design and finalize

FIGURE 2.1
DFSS IDDOV Activities for Process and Product Design.

Identify	Define	Design	Optimize	Validate
• Project charter • Stakeholder analysis • Project plan • Risk matrix • Responsibilities matrix • Items for resolution (IFR) • Ground rules • Communication plan	• Critical to satisfaction (CTS) summary and targets • Data collection plan • Voice of the customer (VOC) • Quality function deployment (QFD) • Benchmarking • Operational definitions • Interviewing • Focus groups • Surveys • Affinity diagram • Market research • Strength–weakness–opportunity–threat (SWOT) analysis • Kano analysis	• Process element summary • Process map • Basic statistics • Failure mode and effect analysis • Risk assessment • Simulation • Prototyping • Design of experiments • Process analysis • Multivoting • Criteria-based matrix • Pugh concept selection technique • Process analysis • Waste analysis • Voice of process (VOP) matrix	• Process capability • Simulation • Implementation plan • Process map • Communication plan • Process analysis • Waste analysis • Cost/benefit analysis • Statistical process control • Training plans • Procedures • Mistake • Proofing • Design of experiments • Pilot	• Prototyping • Testing • Pilot • Mistake proofing • Dashboards • Scorecards • Statistical process control • Statistical analysis • Hypothesis tests • Analysis of variance (ANOVA) • Design of experiments • Replication opportunities

FIGURE 2.2

DFSS IDDOV Tools and Deliverables for Process and Product Design.

	Identify Activities	Tools/Deliverables
1	• Develop project charter	• Project charter • Risk matrix
2	• Perform stakeholder analysis	• Stakeholder analysis definition • Stakeholder commitment scale • Communication plan
3	• Develop project plan	• Project plan • Responsibilities matrix • Items for resolution (IFR) • Ground rules

FIGURE 2.3
Identify Phase Activities and Tools/Deliverables.

2.2 Identify Phase

The purpose of the Identify phase is to define the business problem or opportunity, to scope the project by developing a project charter, and to identify the stakeholders impacted by the project. The main activities to be performed in the Identify phase are as follows:

1. Develop project charter
2. Perform stakeholder analysis
3. Develop project plan

Figure 2.3 shows the main activities mapped to the tools or deliverables most typically used during these activities.

2.2.1 Develop Project Charter

The tools applied in the Identify phase of the IDDOV are the same as those used in the DMAIC Define phase. The team structure including Black Belt and Master Black Belt mentors, project champions and sponsors, process owners, and working team members would be utilized in the DFSS IDDOV methodology similar to the Six Sigma DMAIC methodology. The project charter elements would be similar, except the scope can be somewhat more difficult to define because we do not necessarily have an existing process or product to use as a scope, or a process that can be documented using a SIPOC (Suppliers-Input-Process-Output-Customers) diagram that helps to define the process, inputs, and outputs, as well as the stakeholders of the process. However, thinking through what the potential process steps would be and who would supply inputs and transform these into outputs, and who would receive those outputs would still be helpful to conceptually consider.

> *Project Name*: Name of the Design for Six Sigma Project
>
> *Project Overview*: Background of the project
>
> *Problem Statement*: Business problem, describe what, when, impact, consequences
>
> *Customer/Stakeholders*: (Internal/External) Key groups impacted by the project
>
> *What Is Important to These Customers—Critical to Satisfaction (CTS)*: Critical to satisfaction, the key business drivers
>
> *Goal of the Project*: Describe the improvement goal of the project
>
> *Scope Statement*: The scope of the project, what is in the scope and what is out of scope
>
> *Financial and Other Benefit(s)*: Estimated benefits to the business, tangible and intangible
>
> *Potential Risks*: Risks that could impact the success of the project; can assess risk by probability of occurrence and potential impact to the project
>
> *Milestones*: Identify-Define-Design-Optimize-Validate (IDDOV) phase and estimated completion dates
>
> *Project Resources*: Champion, Black Belt Mentor, Process Owner, Team Members

FIGURE 2.4
Project Charter Template.

The project charter can help to identify the elements that help to scope the project and to identify the project goals.

A project charter template is provided in Figure 2.4.

The following is a description of the elements of the project charter that help to scope and define the business problem:

Project Name: The name of the project. The title of the project should describe the process to be designed or the product to be developed along with the project goal.

Project Overview: This provides a project background and describes the basic assumptions related to your project.

Problem Statement: A clear description of the business problem and motivation/need for the project. What is the challenge or the problem that the business is facing? The problem statement should consider the process that is impacted or the problem that a new product would solve. Define the measurable impact of the problem. The team should be specific as to what is happening, when it is occurring, and what the impact or consequences are to the business of the existing problem.

Customers/Stakeholders: Define the customers, both internal and external, and the stakeholders that are being impacted by the problem, process, or product.

Critical to Satisfaction: Identify what is important to each customer/stakeholder group. They can be identified by what is critical to quality (defects), delivery (time), and cost.

Goal of the Project: What is the quantifiable goal of the project? It may be too early in the problem-solving method to identify a clear target, but at least a placeholder should be identified relating to what should be measured and assessed in order to meet the customers' requirements.

Scope Statement: The scope should clearly identify the process or the product to be designed and what is included or excluded from the scope for the Design for Six Sigma project. The scope can also address the organizational boundaries to be included and possibly, more importantly, which should be excluded. It can also include a temporal scope of the timing of the project and data collection activities. The deliverable scope includes what specifics should be delivered from the project, such as the product to be developed or the process to be designed.

Projected Financial and Other Benefits: This describes potential savings, revenue growth, cost avoidance, cost reduction, cost of poor quality (COPQ), as well as less tangible benefits such as impact to morale, elimination of waste, and inefficiencies.

Potential Risks: Brainstorm the potential risks that could impact the success of the project. Identify the probability that the risk could occur on a high, medium, or low scale. Identify the potential impact to the project if the risk does occur on a high, medium, or low scale. The risk mitigation strategy identifies how the team would potentially mitigate the impact of the potential risk if it does occur.

Project Resources: Identify the project leader who is in charge of the overall project. Identify the division and department of the project leader or project team. Identify the process or product owner, the person who is ultimately responsible for implementing process or product. The project champion is typically at the director (or above) level and has the authority to remove barriers to successful project implementation. The project sponsor is the executive-level person who sponsors the project initiative and is the visible representative of the project and the improvements. The Continuous Improvement Mentor or the Master Black Belt is the team's coach who helps mentor the team members in application of the tools and the IDDOV methodology. Finance is the financial representative who approves the financial benefits or savings established during the project. Team members or support resources are those people who are part of the project team or who provide support, information, or data to the project team.

Milestones: The milestones are the estimated key dates when each phase will be completed and when the project improvements will be approved.

2.2.2 Perform Stakeholder Analysis

It is critical to clearly identify the customers and stakeholders impacted by the project because the quality of the process or product is defined by the customers. Quality is measured by first understanding and then exceeding the customers' requirements and expectations. There is a high cost of an unhappy customer: 96% of unhappy customers never complain; 90% of those who are dissatisfied will not buy again; and each unhappy customer will tell his or her story to as many as 14 other people (Pyzdek, 2003).

Customers and stakeholders can be peers, people who report to me, my boss, other groups within the organization, suppliers, and external customers. Customers can include both internal and external customers of the process. Each process does not always interface directly with an external customer of the company but will have internal customers. Internal customers are people who receive some output from the process, such as information, materials, product, or a service step. It is ultimately the boundary of the process that is being designed or who will use or purchase the product that determines who the customer is.

The stakeholder analysis definition identifies the stakeholder groups, their role, and how they are impacted as well as their concerns related to the process. There is an additional column in the definition matrix which provides a quick view of whether the impact is positive (+), such as reducing variation, or negative (−), such as resistant to change. Figure 2.5 is an example of a stakeholder definition.

The next step in the stakeholder analysis is to understand the stakeholders' attitudes toward change, as well as potential reasons for resistance. Additionally, the team should understand the barriers to change as a result of the resistance. Next, the respective activities, plans, and actions should be developed that can help the team overcome the resistance and barriers to change. A definition of how and when each stakeholder group should participate in the change effort should be developed in the Define phase and then updated throughout the IDDOV project. Figure 2.6 shows a stakeholder commitment analysis. The stakeholder commitment scale can be used to summarize where the stakeholders are regarding their acceptance or resistance to change. The team should determine, based on initial interviews and prior knowledge of the stakeholder groups, the current level of support or

Stakeholders	Role Description	Impact/Concern	+/−
External customer	Customers who receive our marketing efforts related to marketing programs, including advertising circulars and commercials	• Timely information • Accurate information • Coupons	+ + +
Marketing	Internal marketing department that plans, develops, and deploys marketing programs	• Timely deployment • Ability to reach and impact customers	+ +
Information Technology	Information technology department that provides technology	• Clear requirements • Accurate data	+ +

FIGURE 2.5
Stakeholder Analysis Definition.

resistance to the project. Strongly supportive indicates that these stakeholders are supportive of advocating for and making change happen. Moderately supportive indicates that the stakeholders will help, but not strongly. The neutral stakeholders will allow the change and not stand in the way but will not go out of their way to advocate for the change. Moderately against stakeholders do not comply with the change and have some resistance to the project. Strongly against stakeholders will not comply with the change and will actively and vocally lobby against the change. A strategy to move the stakeholders from their current state to where the team needs them to be by the end of the project should be developed. This change strategy should include how the team will communicate with the stakeholders and any activities in their action plan to gain support and implement change.

2.2.3 Develop Project Plan

The project planning activities would include developing a detailed work breakdown structure (WBS) and project plan with roles, responsibilities, estimated durations, and prerequisite relationships of the activities. A responsibilities matrix identifies who is responsible for what during the project and is an important part of the Identify phase to clearly set the expectations of the team members. A sample responsibilities matrix is shown in Figure 2.7. The ground rules also help to clarify expectations of behavior and how the team will operate. A communication plan can help to clearly identify how the team will communicate and interact with the stakeholders. A risk plan, often part of the project charter, can be used to identify potential risks that could impede project progress, as well as identify mitigation and control strategies to avoid and control the risks should they occur. A sample project plan for the IDDOV methodology is shown in Figure 2.8. Additional tasks can be identified within each phase and major activity. Excel® or a

Stakeholders	Stakeholder Commitment Scale					Communication Plan	Action Plan
	Strongly Against	Moderate Against	Neutral	Moderate Support	Strongly Support		
External customer		X →			O	• Surveys • Market research • Test pilot	
Marketing			X →		O	• E-mail • Meetings	• Engage on project
Information technology	X →			O		• Intranet • E-mail • Meetings	• Engage on project • Communicate process and requirements

Note: X, at beginning of project, O, at end of project.

FIGURE 2.6
Stakeholder Commitment Scale.

Role Responsibility	Team Leader	Black Belt	Champion	Process Owner	Team Members
Facilitate meetings	X				
Manage project	X				
Mentor team members	X	X			
Transfer knowledge of Design for Six Sigma tools		X			
Remove roadblocks			X		
Monitor project progress			X		
Approve project			X		
Implement improvements				X	
Subject matter expertise				X	
Apply Design for Six Sigma tools					X
Statistical analysis					X
Data collection					X

FIGURE 2.7
Responsibility Matrix.

project-planning software such as Microsoft® Project can be used to track tasks completed against the project plan. An important part of the project planning is to perform a risk analysis to identify potential risks that could impact the successful completion of the project. The team can brainstorm potential risks to the project. They can also assess the probability that each risk would occur on a scale of high, medium, or low occurrence. The impact of the risk should also be assessed—that is, if the risk were to occur, what level of impact would it have on the successful completion of the project (high, medium, or low)? It is also important to develop a risk mitigation strategy that identifies, if the risk occurs, how the team will mitigate the impact of the risk to reduce or eliminate the impact of the risk. Figure 2.9 shows a simple risk matrix.

Another tool that is useful while planning and managing the project is an item for resolution (IFR) form. This helps the team to document and track items that need to be resolved. It enables the team to complete the planned agendas in meetings by allowing a place to "park" items that arise that cannot be resolved in the meeting, due to time constraints, lack of data, or access to appropriate decision makers. Figure 2.10 shows an IFR form. It includes a description of the item to be resolved. A priority (high, medium, low) should be assigned to each item. The status of the item, open (newly opened), closed (resolved), hold (on hold—not being actively worked) should be identified. The owner who is responsible for resolving the issue, as well as the dates

Activity Number	Phase/Activity	Duration	Predecessor	Resources
1.0	Identify			
1.1	Develop project charter			
1.2	Perform stakeholder analysis		1.1	
1.3	Develop project plan		1.2	
2.0	Define		1.0	
2.1	Collect voice of the customer (VOC)			
2.2	Identify critical to satisfaction (CTS) measures and targets		2.1	
2.3	Translate VOC into technical requirements		2.2	
3.0	Design		2.0	
3.1	Identify process elements			
3.2	Design process		3.1	
3.3	Identify potential risks and inefficiencies		3.2	
4.0	Optimize		3.0	
4.1	Implement process			
4.2	Assess process capabilities		4.1	
4.3	Optimize design		4.2	
5.0	Validate		4.0	
5.1	Validate process			
5.2	Assess performance, failure modes, and risks		5.1	
5.3	Iterate design and finalize		5.2	

FIGURE 2.8
Project Plan.

Potential Risks	Probability of Risk Occurring (High/Medium/Low)	Impact of Risk (High/Medium/Low	Risk Mitigation Strategy

FIGURE 2.9
Risk Matrix.

Number (#)	Issue	Priority	Status	Owner	Open Date	Resolved Date	Resolution

FIGURE 2.10
Item for Resolution (IFR) Form.

that the item was opened and resolved, should be completed on the IFR form. A description of the resolution should also be included. This helps the team keep track of key decisions and ensures that the items are resolved to the satisfaction of all team members. The log of IFRs can also be used during the lessons learned activity after the project is complete to identify where the problems arose and how they were resolved, to incorporate these items into the risk mitigation strategies for follow-on projects.

Another helpful tool that can be developed in the Identify phase, but should be used throughout the DFSS project, is a communication plan. The communication plan can be used to identify strategies for how the team will communicate with all key stakeholders. It can be useful to help overcome resistance to change by planning how frequently and in the manner in which the team will communicate with the stakeholders. Each key stakeholder or audience of a communicated message should be identified. The objectives or message that will be communicated is then developed. The media or mechanism of how to communicate with the audience is then identified. This can be face-to-face, e-mail, Web sites, and so forth. The frequency of the communication is important, especially for those more resistant to change having more frequent communication. The last element of the communication plan is to clearly identify who is responsible for developing and delivering the communication to the audience. The communication plan is shown in Figure 2.11.

2.2.3.1 Team Meeting Management

Some best practices for team meeting management are as follows:

- Respect people and their time.
- Determine critical/required participants for e-mails, meetings, and decisions.

Audience	Objectives/Message	Media/Mechanism	Frequency	Responsible

FIGURE 2.11
Communication Plan.

- Cancel or schedule meetings ahead of time.
- Always create a meeting agenda and send it out in advance of the meeting. The agenda should include required and optional participants.
- Recap action items and meeting minutes.
- Use voting in e-mails to make easy decisions, or agree upon a meeting time.
- Track meeting attendance, and resolve habitual lack of attendance.

The planned meeting agenda should include the following:

- Date, time, and proposed length of the meeting
- Meeting facilitator's name
- Meeting location
- Required and optional attendees
- Purpose of the meeting
- Desired outcomes
- Topic with time and proposed outcome for each topic

Some tips that the meeting facilitator can use to keep the meeting productive are as follows (Scholtes, Joiner, and Streibel, 2003):

- Listen and then restate what you think you heard.
- Ask for clarification and examples.
- Encourage equal participation and circle the group.
- Summarize ideas and discussion.
- Corral digressions and get back to the agenda.
- Close the discussion.

2.2.4 Summary

The Identify phase is a critical phase of the project. It is important to spend ample time in this phase developing the project charter and getting the buy-in of the project champion, the team members, and all stakeholders. The time spent clearly defining the scope of the project will reap dividends by reducing issues during the remaining phases of the project. A process or a problem poorly defined will require the team to revisit the project charter and definition activities when the project bogs down or loses focus in subsequent phases.

2.3 Define Phase

The purpose of the Define phase is to understand the voice of the customer, what is important to the customer as defined by the CTS (critical to satisfaction) characteristics, and to translate the customer's requirements into the technical elements of the process or product to be designed. The following activities can be applied to meet the objectives of the Define phase:

1. Collect voice of the customer (VOC)
2. Identify CTS measures and targets
3. Translate VOC into technical requirements

Figure 2.12 shows the main activities mapped to the tools or deliverables most typically used during that step.

2.3.1 Collect Voice of the Customer (VOC) and Identify Critical to Satisfaction (CTS) Measure and Targets

The voice of the customer information should be collected to define the customers' expectations and requirements with respect to the service delivery

	Define Activities	Tools/Deliverables
4	• Collect voice of the customer (VOC)	• Data collection plan • VOC • Interviewing, surveying, focus groups, market research
5	• Identify critical to satisfaction (CTS) measures and targets	• Critical to satisfaction (CTS) summary and targets • Affinity diagram • Quality function deployment (QFD) • Operational definitions • Strength-weakness-opportunity-threat (SWOT) analysis
6	• Translate VOC into technical requirements	• QFD • Benchmarking • Kano analysis

FIGURE 2.12
Define Phase Activities and Tools/Deliverables.

process or product requirements. VOC is an expression for listening to the external customer and understanding his or her requirements for your product or service. Examples of requirements are expectations for responsiveness, such as turnaround time on vendor (customer) invoices, or error rates, such as employee (customer) expectations of no errors on their paychecks. The voice of the customer can be captured through interviewing, surveys, focus groups with the customers, complaint cards, warranty information, and competitive shopping. Personal interviews are an effective way to gain the voice of the customer; however, it can be expensive, and training of interviewers is important to avoid interviewer bias. Also, additional questioning may be necessary to eliminate misunderstandings and clarify customer requirements. The objectives of the interview should be clearly defined before the interviews are held.

A data collection plan (Figure 2.13) could be used to identify the data to be collected that would support the assessment of the proposed CTS and to validate these CTS from the customers' perspective.

The data collection plan ensures the following:

- Measurement of critical to satisfaction metrics
- Identification of the right mechanisms to perform the data collection
- Collection and analysis of data
- Definition of how and who is responsible to collect the data

Critical to Satisfaction (CTS)	Metric	Data Collection Mechanism (Survey, Interview, Focus Group, etc.)	Analysis Mechanism (Statistics, Statistical Tests, etc.)	Sampling Plan (Sample Size, Sample Frequency)	Sampling Instructions (Who, Where, When, How)
Speed to market	Cycle time	Project management tool	Statistics (mean, variance); t-test	One year of projects	Collect data from project management system for last year
	Functionality delivered	Requirements traceability tool	Count	50 projects (30 development, 20 support)	Extract data based on sampling plan

FIGURE 2.13
Data Collection Plan for Software Application Development Six Sigma Project.

The steps for creating a data collection plan are as follows:

1. Define the critical to satisfaction metrics.
2. Develop metrics.
3. Identify data collection mechanism(s).
4. Identify analysis mechanism(s).
5. Develop sampling plans.
6. Develop sampling instructions.

A description of each step in the data collection plan development follows.

2.3.1.1 Define the Critical to Satisfaction Criteria

Critical to satisfaction is a characteristic of a product or service which fulfills a critical customer requirement or a customer process requirement. CTS characteristics are the basic elements to be used in driving process measurement, improvement, and control (George et al., 2005).

2.3.1.2 Develop Metrics

In this step, metrics are identified that help to measure and assess improvement related to the identified CTS. Some rules of thumb for selecting metrics are to (Evans and Lindsey, 2002)

- Consider the vital few versus the trivial many.
- Metrics should focus on the past, the present, and the future.
- Metrics should be linked to meet the needs of shareholders, customers, and employees.

It is vital to develop an operational definition for each metric, so it is clearly understood how the data will be collected by anyone who collects it. The operational definition should include a clear description of a measurement, including the process of collection. Include the purpose and metric measurement. It should identify what to measure, how to measure it, and how the consistency of the measure will be ensured. A summary of an operational definition follows.

2.3.1.2.1 Operational Definition

Defining the Measure, Definition: A clear, concise description of a measurement and the process by which it is to be collected

Purpose: Provides the meaning of the operational definition, to provide a common understanding of how it will be measured

Clear Way to Measure the Process:

Identifies what to measure

Identifies how to measure

Makes sure the measuring is consistent

2.3.1.3 Identify Data Collection Mechanism(s)

Next you can identify how you will collect the data for the metrics. Some data collection mechanisms include customer surveys, observation, work sampling, time studies, customer complaint data, e-mails, Web sites, and focus groups.

2.3.1.3.1 Customer Surveys

Interviews, focus groups, surveys, and market research are some of the most common ways to collect the VOC. The main difference between DFSS and Six Sigma would be that existing customer complaints, warranty information, and other data from an existing process or product would not be available or could not apply to our new process or product that we are designing. However, data from similar processes or products may be used in these cases. For redesign of processes and products, then historical information would be available for a DFSS project.

Customer surveys are a typical way to collect VOC data. The response rate on surveys tends to be low, 20% is a "good" response rate. It can also be extremely difficult to develop a survey that avoids bias and asks the questions that are desired. Customer survey collection can be quite expensive. The steps to create a customer survey are as follows (Malone, 2005):

1. *Conceptualization*: Identify the survey objective and develop the concept of the survey, and what questions you are trying to answer from the survey.
2. *Construction*: Develop the survey questions. A focus group can be used to develop and/or test the questions to see if they are easily understood.
3. *Pilot (Trial Survey)*: Pilot the questions by having a focus group of representative people from your population. You would have them review the questions, identify any unclear or confusing questions, and tell you what they think the questions are asking. You would not use the data collected during the pilot in the actual results of the surveys.

2.3.1.4 Item Analysis

Item analysis provides a statistical analysis to determine which questions answer the same objectives, as a way to reduce the number of questions. It is important to minimize the number of questions and the total time required to take the survey. Typically, the survey time should be 10 minutes or less to help ensure participants complete the entire survey.

2.3.1.4.1 Revision

Revise the survey questions, and roll out the customer survey, or pilot again if necessary.

2.3.1.4.2 *Focus Groups*

Focus groups are an effective way to collect VOC data. A small representative group, typically 7 to 10 people are brought together and asked to respond to predetermined questions. The focus group objective should be developed, and the questions should support the objective. The participants should be selected by a common set of characteristics. The goal of a focus group is to gather a common set of themes related to the focus group objective. There is no set sample size for focus groups. Multiple focus groups are typically run until no additional themes are derived. Advantages of focus groups are the following:

- They tend to have good face validity, meaning that the responses are in the words of the focus group participants.
- Typically more comments are derived than in an interview with one person at a time.
- The facilitator can probe for additional information and clarification.
- Information is obtained relatively inexpensively.

Some of the disadvantages of focus groups are as follows:

- The skills of the facilitator dictate the quality of the responses.
- They can be difficult to schedule.
- It can be difficult to analyze the dialogue due to participant interactions.

Affinity diagrams organize interview, survey, and focus group data after collection. The affinity diagram organizes the data into themes or categories. The themes can first be generated and then the data can be organized into the themes, or the detailed data can be grouped into the themes. An example of a simple affinity diagram for ways to study for a Six Sigma Black Belt exam is shown in Figure 2.14.

Affinity Diagram for Six Sigma Black Belt Exam

Resources	Preparation	Motivation
• Tab training materials	• Study material	• Schedule exam
• Get other references	• Apply tools on projects	• Motivate self
• Review LSS Pocket Toolbook	• Study in groups	• Talk to other candidates
• Discuss with mentor	• Attend refresher	

FIGURE 2.14
Affinity Diagram.

It is important to clearly summarize the CTS so that we can operationally define the metrics and then translate these into the process elements that form the technical requirements of our new process.

2.3.1.4.3 Benchmarking

Benchmarking is a tool that provides a review of best practices to be potentially applied to enhance the designed processes or improve the design of your product. Benchmarking can be powerful in the DFSS arena so that the organization looks outside of itself to understand the industry and even best practices outside of the industry that can be used as a model for our process or product design. Process benchmarking can be performed to understand best practice processes. The organization should document the process that they will benchmark and then select who they will benchmark. It is not necessary to benchmark a company in the same industry, but to focus on the process to be benchmarked and select an organization that is known for having world-class or best-practice processes. The next step is to work with the organization to collect the data and understand how the data can be used to identify ways to enhance your processes. This is similar to Motorola's benchmarking process (Evans and Lindsey, 2002). It is important to be careful when process benchmarking to ensure that you are comparing apples to apples, meaning that the organization's characteristics are similar to your own, so that the benchmarked process applies to your process. Competitive product benchmarking can be performed if you are designing a product. This is assessing how your product concept compares to similar competitive products, based on important customer requirements.

2.3.2 Translate VOC into Technical Requirements

2.3.2.1 Quality Function Deployment

Quality function deployment (QFD) and the House of Quality (HOQ) is an excellent tool to help to translate customer requirements from the voice of the customer into the technical requirements of your product, process, or service. The customer requirements would be prioritized by the customers through market research techniques. The strength of the relationship between the customer requirements and the technical requirements would be identified by the process design team. These relationship strengths would be multiplied by the CTS priorities to derive a relative weighting of the technical requirements. For a product- or system-level HOQ, you should only use the VOC that are new needs (features or performance the product has not historically delivered); unique needs (features or performance that are distinctive or highly desired beyond the numerous other less-demanding needs that must be provided); and difficult-to-fulfill needs (features or performance that are highly desired but are quite difficult for your business to develop and will require special efforts/investments of resources on your part) (Paryani, Masoudi, and Cudney, 2010). A QFD format is shown in Figure 2.15.

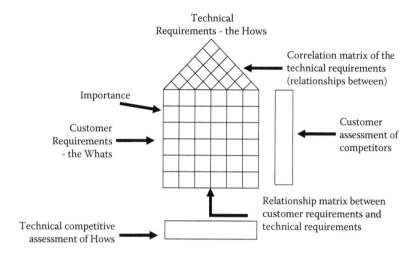

FIGURE 2.15
Quality function deployment House of Quality.

QFD can lead to (Paryani et al., 2010)

- Better understanding of customer requirements
- Increased customer satisfaction
- Reduced time to market and lower development costs
- Structured integration of competitive benchmarking into the design process
- Increased ability to create innovative design solutions
- Enhanced capability to identify those specific design aspects that have the greatest overall impact on customer satisfaction
- Better teamwork in cross-functional design teams
- Better documentation of key design solutions

The steps for creating a HOQ are as follows (Evans and Lindsey, 2002):

1. Define the customer requirements or CTS characteristics from the VOC data. The customer can provide an importance rating for each CTS.
2. Develop the technical requirements with the organization's design team.
3. Perform a competitive analysis, having the customers rank your product, process, or service against each CTS characteristic to each of your competitors.
4. Develop the relationship correlation matrix by identifying the strength of relationship between each CTS and each technical

requirement. Typically there is a numerical scale of 9 (high strength of relationship), 3 (medium strength of relationship), 1 (low strength of relationship), and blank (no relationship).

5. Develop the trade-offs or relationships between the technical requirements in the roof of the HOQ. You can identify a positive (+) relationship between the technical requirements, as one requirement increases the other also increases; with no relationship (blank) or a negative (–) relationship, there is an inverse relationship between the two technical requirements. An example of a positive relationship can be illustrated in the design of a fishing pole. The line gauge and tensile strength both increase as the other increases. A negative relationship can be illustrated by line buoyancy and tensile strength. As tensile strength of the line increases, the buoyancy will be less.

6. The priorities of the technical requirements can be summarized by multiplying the importance weightings of the customer requirements by the strength of the relationships in the correlation matrix. This helps to identify which of the technical requirements should be incorporated into the design of the product, process, or service first. We will use the identified technical requirements as the process elements for our process or as technical product requirements for a product as inputs to the Design phase.

2.3.2.2 Interpreting the House of Quality

The House of Quality is interpreted as follows (Paryani et al., 2010):

- Blank columns are unnecessary design requirements.
- Blank rows are missed customer requirements.
- Rows with no strong relationships are missed customer requirements.
- Identically weighted rows indicate a possible misunderstanding of customer requirements.
- Strong diagonal patterns indicate that customer requirements probably contain design solutions.
- Complete or nearly complete row shows that customer requirement involves a cost, reliability, or safety problem.
- Complete or nearly complete column is a design requirement that involves a cost, reliability, or safety problem.
- A large number of weak relationships indicate too much fuzzy weight, clear decisions will be confounded with noise.

2.3.2.3 Kano Analysis

The Kano model defines three types of quality requirements (Cudney, Elrod, and Uppalanchi, 2012):

- *One-dimensional quality*: The one-dimensional or linear quality is a specifically requested item. It is a stated want identified on a customer survey. If present, the customer is satisfied. If this characteristic is absent, the customer is dissatisfied. An example is asking for a rare steak.

- *Expected or basic quality*: These elements or customer requirements are not specifically requested, but they are assumed by the customer to be present. If they are present the customer is neither satisfied nor dissatisfied. If they are absent, the customer would be very dissatisfied. An example is clean silverware in a restaurant.

- *Exciting or delightful quality*: This level of quality characteristics is unknown to the customer. It is not something that they would think to ask for. These elements are the most difficult to define and develop. If present, the customer is very satisfied. If absent, the customer is neither satisfied nor dissatisfied. An example is fine linen or fresh floral arrangements.

The Kano model also defines how the achievement of these requirements affects customer satisfaction. The Kano model is shown in Figure 2.16.

2.4 Design Phase

The purpose of the Design phase is to understand the elements of the process or product which can ensure the CTS of the customers and stakeholders are met; to design the new process or product; and to identify potential risks,

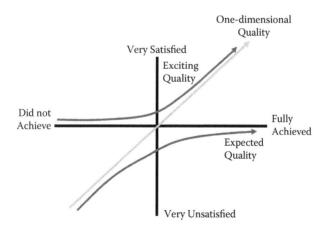

FIGURE 2.16
Kano model.

	Design Activities	Tools/Deliverables
7	• Identify process elements	• Process element summary
8	• Design process	• Basic statistics • Simulation • Prototyping • Design of experiments (DOE) • Process analysis • Multivoting • Criteria-based matrix • Pugh concept selection technique • Voice of process (VOP) matrix
9	• Identify potential risks and inefficiencies	• Failure mode and effect analysis (FMEA) • Risk assessment • Process analysis • Waste analysis

FIGURE 2.17
Design Phase Activities and Tools/Deliverables.

failures, and inefficiencies that could occur in the new process. Following are the main activities to be performed in the Design phase:

1. Identify process elements.
2. Design process.
3. Identify potential risks and inefficiencies.

Figure 2.17 shows the main activities mapped to the tools or deliverables most typically used during the Design phase.

2.4.1 Identify Process Elements

The first step in the Design phase is to analyze the VOC data that were collected in the Define phase. Attribute survey analysis using chi-square statistical analysis would be used to analyze attribute survey data. The results and data collected from the VOC would be used to generate the elements that would be incorporated into a process or product, or potential alternate process or product concepts. Potential elements could be categorized by people, process, and technology. The people aspects would be which organizations and roles would be involved in owning and contributing to the process; the cultural and political aspects, resistance to change, training and skill sets available, and organizational structure. The process elements could pertain to any policies and procedures that may impact the process, understanding the activities needed to be performed as well as how to measure and assess performance. The technology elements would pertain to what technologies would be needed, such as using a SharePoint site, or perhaps an off-the-shelf or internally developed information system.

There are many techniques that are part of the DFSS tool kit that can help to generate and brainstorm process elements and concepts, such as traditional brainstorming and Nominal Group Technique, channel and analogy brainstorming, antisolution brainstorming and brainwriting, assumption busting, and the Theory for Inventive Problem Solving (TRIZ) (Chowdhury, 2002).

Traditional brainstorming includes sharing ideas in a group and writing them on a flip chart or whiteboard. Nominal Group Technique structures the brainstorming into first a silent generation then a round-robin idea sharing. Important in any brainstorming activity is to hold the criticism and evaluation until after ideas are generated. Channel brainstorming allows a group to focus on a subcategory of a task to make the brainstorming more manageable. Analogy brainstorming allows participants to focus on a similar or parallel issue to generate ideas, and then link it back to the original issue. Antisolution brainstorming asks the participants to generate ideas of how they could make the process even worse, punching holes in your own argument. In brainwriting, each participant writes down an idea and then passes it to the person next to them who then builds on the idea or concept. Assumption busting is when the brainstorming group, instead of asking "why?" asks "why not?"

TRIZ (The Theory for Inventive Problem Solving), pronounced "trees," was developed by Genrich Altshuller. He developed the TRIZ principles by reviewing approximately 40,000 patent applications and extracting the key principles. He developed a set of principles that can be used to cultivate inventions to eliminate corporate contradictions and problems while generating creative solutions. A TRIZ principle encourages the team to look at the past, present, and future of the process when designing the process. The following steps describe the TRIZ process:

1. Think of the ideal vision, process, or system.
2. Think of ways to improve the process or function.
3. Think of ways to eliminate or reduce undesired functions.
4. Think of ways to segment the process.
5. Think of ways to copy existing ideas or processes.
6. Think of a disposable concept.

There is a great deal of depth and richness in the TRIZ concept, related to a tangible product design. Presented here are the elements of TRIZ that could apply to designing intangible service processes.

2.4.2 Design Process

The Pugh Concept Selection technique is a technique for evaluating and selecting concepts. If you have several different process elements or product concepts to choose from, you could use this technique. You would first brainstorm potential solutions or concepts and then generate criteria upon which to compare the concepts. Then you would select one of the concepts as the "candidate" concept. It does not matter which concept you select as the candidate concept; however, for a redesign of a process or product you could select the current process or product. You then compare each of the other (new) concepts to the candidate for each comparison criteria. If the new concept is better than the candidate for those criteria, you would place a plus sign (+) in the cell where the new concept intersects the criteria. If the new concept is worse than the candidate concept for the criteria, a minus sign (–) is placed in the cell. If the new concept is the same as the candidate on those criteria, a zero (0) or *S* for same is placed in the cell. Figure 2.18 shows a Pugh's Concept Selection matrix. You would select the few concepts with the most pluses and the fewest minuses. You could also attack the weaknesses of the few concepts and enhance them with the strengths of the surviving alternatives.

After you identify the process or product elements or concepts, the team can then design the process or product. A process map is a great tool to communicate the steps of the new process. It helps to think through sequencing, who does what in the process, as well as the information that is needed to perform each step of the process and what output is transformed by each process step.

The elements with the pluses can be used to form a hybrid product or process that takes the best aspects of each concept and combines them to form

		Concepts						
Criteria	1	2	3	4	5	6	7
A	–	–	–	0	Candidate	0	–
B	–	0	–	–	Concept	0	–
C	+	+	–	–		–	–
D	+	–	–	+		–	+
E	+	+	–	–		–	–
Pluses	3	2	0	1		0	1
Minuses	2	2	5	3		3	4
Zeros	0	1	0	1		2	0

FIGURE 2.18
Pugh's Concept Selection Technique.

a superior concept. This may result in a shortened list of concepts. The team can then use the Pugh's Concept Selection matrix to further analyze the elements and select the best concept or further incorporate the best elements.

2.4.3 Identify Potential Risks and Inefficiencies

2.4.3.1 Process Analysis

Process and waste analyses can be performed to identify potential process inefficiencies and wasteful activities. The process analysis helps to identify which activities you have defined in the process that do not add value, which could be further eliminated, combined, or reduced. The waste analysis identifies activities that do not add value and are wasteful.

To identify inefficient process activities in your newly designed process or in the manufacturing processes identified to develop your product, the team can perform a process analysis coupled with waste elimination. A process analysis consists of the following steps:

1. Document the process using process maps.
2. Identify non-value-added activities and waste.
3. Consider eliminating non-value-added activities and waste.
4. Identify and validate (collect more data if necessary) root causes of non-value-added activities and waste.
5. Begin generating improvement opportunities.

Value-added activities are those activities that the customer would pay for and that add value for the customer. Non-value-added activities are those that the customer would not want to pay for or do not add value for the customer. Some are necessary, such as for legal, financial reporting, and documentation reasons that are known as necessary non-value-added activities and may be reduced; however, others are unnecessary and should be reduced or eliminated. You can assess the percent of value-added activities as

$$\frac{\text{Number of value-added activities}}{\text{Number of total activities}} \times 100\%$$

where value-added activities include operations that add value for the customer and non-value-added activities include delays, storage of materials, movement of materials, and inspections. The number of total activities includes the value-added activities and the non-value-added activities.

You can also calculate the percent of value-added time as

$$\frac{\text{Total time spent in value-added activities}}{\text{Total time for process}} \times 100\%$$

Typically, the percent of value-added time is about 1% to 5%, with total non-value-added time equal to 95% to 99%.

During the process analysis, the team can focus on areas to identify inefficiencies in the following areas:

- Can labor-intensive process be reduced, eliminated, or combined?
- Can delays be eliminated?
- Are all reviews and approvals necessary and value added?
- Are decisions necessary?
- Why is rework required?
- Is all of the documentation, tracking, and reporting necessary?
- Are there duplicated processes across the organization?
- What is slipping through the cracks causing customer dissatisfaction?
- What activities require accessing multiple information systems?
- What requires excess travel? Look at the layout requiring the travel.
- Is it necessary to store and retrieve all of that information? Do we need that many copies?
- Are the inspections necessary?
- Is the sequence of activities or flow logical?
- Are standardization, training, and documentation needed?
- Are all of the inputs and outputs of a process necessary?
- How are the data and information stored and used?
- Are systems slow?
- Are systems usable?
- Are systems user friendly?
- Can you combine tasks?
- Is the responsible person at too high or low of a level?

2.4.3.2 Waste Analysis

Waste analysis is a Lean tool that distributes waste into eight different categories to help brainstorm and eliminate different types of wastes. The eight wastes are all considered non-value-added activities and should be reduced or eliminated when possible. Waste is defined as anything that adds cost to the product without adding value. The eight wastes are

- *Transportation*: Moving people, equipment, materials, and tools
- *Overproduction*: Producing more product or material than is necessary to satisfy the customers' orders or faster than is needed

- *Motion*: Unnecessary motion, usually at a micro or workplace level
- *Defects*: Any errors in not making the product or delivering the service correctly the first time
- *Delay*: Wait or delay for equipment or people
- *Inventory*: Storing product or materials
- *Processing*: Effort that adds no value to a product or service; incorporating requirements not requested by the customer
- *People*: Not using people's skills, and mental, creative, and physical abilities

The process metrics that will be embedded in the process should be defined. An operational definition includes the purpose of the measure, as well as a specific and detailed description of how you would measure the metric.

Some other tools, beyond the scope of this text, could include performing simulations, prototypes, and design of experiments to help in designing the process.

2.4.3.3 Failure Mode and Effect Analysis

A failure mode and effect analysis (FMEA) is a great tool to help think through the potential risks in a process or where the product failures can occur. By thinking of potential failure modes for each process step or product component, identifying the probability of occurrence, identifying the impact or severity to the stakeholders if the failures occur, and having the ability to detect the failure, we can develop recommendations to incorporate into the process or product to either reduce the probability for failure, reduce the impact if the failure occurs, and improve the ability to detect the failure. The FMEA is a systemized group of activities intended to recognize and evaluate the potential failure of a product or process, identify actions that could eliminate or reduce the likelihood of the potential failure occurring, and document the entire process.

The FMEA process includes the following steps:

1. Document the process and define functions.
2. Identify potential failure modes.
3. List effects of each failure mode and causes.
4. Quantify effects: severity (SEV), occurrence (OCC), detection (DET).
5. Define controls.
6. Calculate risk and loss.
7. Prioritize failure modes.
8. Take action.
9. Assess results.

A simple FMEA form is shown in Figure 2.19.

Process Step/ Function	Potential Failure Mode	Potential Effects of Failure	Severity (S E V)	Potential Causes of Failure	Occurrence (O C C)	Current Process Controls	Detection (D E T)	Risk Priority Number (R P N)	Recommended Action

FIGURE 2.19
Failure Mode and Effect Analysis Form.

The risk priority number (RPN) is calculated by multiplying the severity times the occurrence times the detection. The severity is estimated for the failure and is given a numerical rating on a scale of 1 (low severity) to 10 (high severity). Occurrence is given a numerical rating on a scale of 1 (low probability of occurrence) to 10 (high probability of occurrence). The detection scale is reversed, where a numerical rating is given on a scale of 1 (failure is easily detected) to 10 (failure is difficult to detect).

A Pareto diagram can be created based on the RPN values to identify the potential failures with the highest RPN values. Recommendations should be developed for the highest-value RPN failures to ensure that they are incorporated into the improvement recommendations for the process or product.

2.5 Optimize Phase

The purpose of the Optimize phase is to understand the elements of the process or product that can ensure the CTS characteristics of the customers and stakeholders are met, to pilot the new process or prototype the new product, to assess process capabilities, and to identify potential risks, failures, and inefficiencies that could occur in the new process. Following are the main activities to be performed in the Optimize phase:

1. Implement pilot process.
2. Assess process capabilities.
3. Optimize design.

Figure 2.20 shows the Optimize activities mapped to the tools and deliverables typically used in the Optimize phase.

2.5.1 Implement Pilot Processes

The team should gain the appropriate approvals to pilot the process or develop a product prototype from the process owners and stakeholders. A presentation of the project to this point may help to communicate the value of the project and the new process or product. To implement the process or develop the product prototype, the team would develop an implementation plan that would include each implementation activity, who would be responsible for implementing each step, the stakeholders the activity would impact, and the due date for when the activity would be complete. Figure 2.21 shows an implementation plan template.

	Optimize Activities	Tools/Deliverables
10	• Implement pilot process	• Implementation plan • Communication plan • Training plan • Procedures
11	• Assess process capabilities	• Process capability • Simulation
12	• Optimize design	• Process map • Process analysis • Waste analysis • Cost/benefit analysis • Statistical process control (SPC) • Mistake proofing • Design of experiments (DOE)

FIGURE 2.20
Optimize Phase Activities and Tools/Deliverables.

2.5.1.1 Statistical Process Control Charts

Statistical process control (SPC) is an effective tool to help ensure your process performance is being attained. SPC charts are a graphical tool for monitoring the activity of an ongoing process. SPC is a methodology that uses the basic graphical and statistical tools to analyze, control, and reduce variability within a process. It is a powerful tool to optimize the information for use in making management decisions. By collecting readily available data from samples, variations in the process, the quality of the end product and service can be detected and corrected before waste is produced.

The most commonly used control charts are also referred to as Shewhart control charts, because Walter A. Shewhart first proposed the general theory in the 1920s at AT&T Western Electric. Figure 2.22 identifies the most commonly used control charts.

These charts can highlight trends and identify when something goes wrong in the process (assignable cause) that would encourage us to investigate the

Activity	Responsible	Due Date	Stakeholders Impacted

FIGURE 2.21
Implementation Plan Template.

Most Common Variables Charts	Most Common Attributes Charts
• X-bar and R-charts (average and range) • X-bar and s-charts (average and standard deviation) • X or IMR (individual and moving range)	• P-charts (proportion nonconforming) • NP-charts (number nonconforming items) • C-charts (number nonconformities) • U-charts (number nonconforming per unit)

FIGURE 2.22
Commonly Used Control Charts.

root cause of the problem. Assignable causes or special cause variations stem from factors that cannot be adequately explained by any single distribution of the process output. These are shifts in output caused by a specific factor such as environmental conditions or process input parameters that affect the process output in unpredictable ways. This type of variation can be detected by simple statistical techniques.

Common causes are sources of variation that produce chance or natural variation. These stem from phenomena constantly active within the system. If common causes of variation are present, the output of the process forms a distribution that is predictable and stable over time. Eliminating common cause variation requires actions on the system.

The process capability would be assessed, by collecting data for the metrics previously identified. For service processes, the data do not necessarily follow a normal distribution; therefore, a nonnormal capability analysis should be used. If attribute control charts are used to control the process, the process capability is the average value or center line of the control chart when the process is in control.

The following steps can be used to implement control charts:

1. Determine the type of chart, quality characteristic, sample size and frequency, and data collection mechanism.
2. Select the rules for out-of-control conditions.
3. Collect the data (10 to 25 subgroups).
4. Order data based on time order.
5. Calculate the trial control limits, create charts (using a statistical program such as Minitab®).
6. Identify out-of-control conditions.
7. Remove points where you can assign causes.
8. Recompute the control limits.

If the process is not meeting the target metrics and expectations of the customers and stakeholders, further redesign of the process can be performed. Further process and waste analysis would be helpful for the redesign. Also, if

Training Topic	Target Audience	Expected Length of Topic	Prerequisite Knowledge	Instructional Strategy
Process mapping	Process analysts	4 hours	Concepts of Six Sigma, Lean, or Design for Six Sigma	Workshop with hands-on exercises building process maps
Design for Six Sigma Identify-Define-Design-Optimize-Validate (IDDOV) methodology	Business analysts, process engineers	12 days across 3 separate weeks, 3 months apart	None	Workshops with hands-on case; mentored work projects

FIGURE 2.23
Training Plan Template.

training was not implemented during the pilot process, it should be considered first to ensure the new process is being consistently understood and practiced, and skill transfer is occurring. Training plans would include the topics to be covered, as well as the targeted training audience, the expected length of the topic, any expected prerequisite knowledge, and the instructional strategies to be applied. Figure 2.23 shows a training template example.

Detailed procedures also help to train stakeholders in the processes to provide the service or to develop the new product to ensure consistency and repeatability of the process. We cannot improve a process if it is not first consistent, stable, and repeatable (i.e., in control). This provides a baseline upon which to further optimize and improve the process.

2.5.2 Assess Process Capabilities

To develop the process capability, you can calculate the DPMO (defects per million opportunities) and related sigma level, or you can calculate the capability indices. We will first discuss DPMO.

2.5.2.1 Defects per Million Opportunities (DPMO)

Six Sigma represents a stretch goal of six standard deviations (σ) from the process mean to the specification limits when the process is centered, but also allows for a 1.5 sigma shift toward either specification limit. This represents a quality level of 3.4 defects per million. This is represented in Figure 2.24.

The greater the number of σ's, the smaller is the variation (the tighter the distribution) around the average. DPMO provides a single measure to compare performance of very different operations, giving an apples-to-apples comparison, not apples to oranges. Figure 2.25 shows a sigma to DPMO conversion incorporating a 1.5 sigma shift over time.

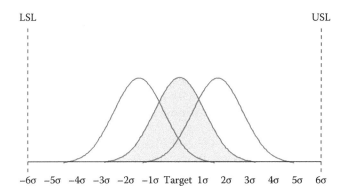

FIGURE 2.24
A 3.4 defects per million opportunities (DPMO) representing a Six Sigma quality level, allowing for a 1.5 sigma shift in the average.

DPMO is calculated as

$$DPMO = \frac{\text{Number of Defects}}{\text{Number of Units} \times \text{Number of Opportunities}} \times 1{,}000{,}000$$

where defects is the number of defects in the sample; units is the number of units in the sample, and opportunities is the number of opportunities for error. For example, for a purchase order with a sample of 100 purchase orders, finding 5 defects, with 30 fields on the purchase order (opportunities for errors), we calculate a DPMO of 1,667 or about 4.4 sigma.

2.5.2.2 Process Capability Study

Process capability is the ability of a process to produce products or provide services capable of meeting the specifications set by the customer or designer.

Sigma Level	Defects per Million Opportunities (DPMO)
6 σ	3.4 DPMO
5 σ	233 DPMO
4 σ	6,210 DPMO
3 σ	66,810 DPMO
2 σ	308,770 DPMO
1 σ	691,462 DPMO

FIGURE 2.25
Sigma to DPMO conversion (assuming 1.5 sigma shift).

You should only conduct a process capability study when the process is in a state of statistical control. Process capability is based on the performance of individual products or services against specifications. According to the central limit theorem, the spread or variation of the individual values will be greater than the spread of the averages of the values. Average values smooth out the highs and lows associated with individuals.

Following are the steps for performing a process capability study:

1. Define the metric or quality characteristic. Perform your process capability study for the metrics that measure your CTS characteristics.
2. Collect data on the process for the metric, take 25 to 50 samples.
3. Perform graphical analysis (histogram).
4. Perform a test for normality.
5. Determine if the process is in control and stable, using control charts. When the process is stable, continue to Step 6.
6. Estimate the process mean and standard deviation.
7. Calculate the capability indices, usually: C_p, C_{pk}

$$Cp = \frac{\text{Upper Specification Limit} - \text{Lower Specification Limit}}{6\sigma}$$

C_{pk} is the minimum of {CPU, CPL}, where

$$CPU = \frac{\text{Upper Specification Limit} - \text{Process Mean}}{3\sigma}$$

$$CPL = \frac{\text{Process Mean} - \text{Lower Specification Limit}}{3\sigma}$$

A process can be in control but *not* necessarily meet the specifications established by the customer or engineering. You can be in control and not capable. You can be out of control or unstable but still meet specifications. There is no relationship between control limits and specification limits. Control limits are calculated based on the voice of the process (i.e., process data). Specification limits are based on the voice of the customer. Your process must be in control before you use the estimates of standard deviation from your process to calculate process capability and your capability indices.

There are typically three scenarios regarding process capability:

1. Process spread is less than the specification spread and within specification limits. The process is quite capable. Figure 2.26 shows this scenario. C_p and C_{pk} are greater than 1.33.

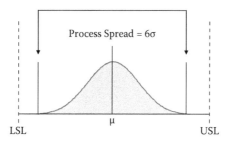

FIGURE 2.26
Process is quite capable.

2. Process spread is equal to specification spread and within specification limits, an acceptable situation, but there is no room for error. If the mean shifts, or variation increases, there will be nonconforming product. Figure 2.27 shows this scenario. $C_p = C_{pk} = 1$.

3. Process spread is greater than the specification spread. The process is *not* capable. Figure 2.28 shows this scenario. C_p and C_{pk} are less than 1.

2.5.3 Optimize Design

To optimize design several tools can be used. Process mapping, process analysis, and waste analysis can be further applied to optimize the newly designed processes and the manufacturing processes that create your new product. A cost/benefit analysis may be necessary to cost justify product or process implementation decisions, especially if new technology or new information systems are required. Statistical process control can be used along with process capability studies to ensure that the processes are stable and that assignable causes are captured and eliminated when discovered.

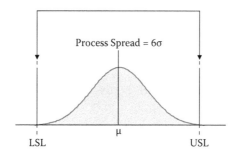

FIGURE 2.27
Process is just capable.

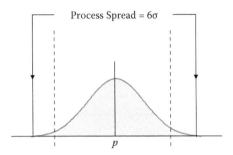

FIGURE 2.28
Process is not capable.

2.5.3.1 Mistake Proofing

Mistake proofing is a tool that helps to prevent errors in your process. Errors are inadvertent, unintentional, accidental mistakes made by people because of the human sensitivity designed into our products and processes.

Mistake proofing, also called Poka Yoke, is the activity of awareness, detection, and prevention of errors that adversely affect our customers and our people and result in waste.

Some of the underlying mistake-proofing concepts are

- You should have to think to do it wrong, instead of right.
- Easy to perform inspection should be done at the source.
- An immediate response should be given when a defect occurs.
- It reduces the need for rework and prevents further work (and cost) on a process step that is already defective.
- Simplifies prevention and repair of defects by placing responsibility on the responsible worker.

2.5.3.2 Design of Experiments

Design of experiments includes a vast array of statistically designed experiments that can be used to identify the factors that impact the quality of the product or processes. Sequential experimentation can be used to first screen potential factors that impact product or process design, and then run more refined studies to identify the levels across the identified critical factors. The following experiments can be applied to help optimize the product or process design:

- Screening experiments define basic, linear, main effects between a Y variable and any number of x values. This approach looks at the effects of independent x variables.

- Main effects and interaction identification experiments define how strong the main effects and the interactions between certain main effects (*x*s) are on the *Y* variables.
- Nonlinear effects experiments use a variety of experimental data gathering and analysis structures to identify and quantify the strength of nonlinear effects of certain *x* variables (second-order effects and beyond as necessary).
- Response surface experiments study relatively small numbers of *x*s for their optimum set points (including the effects due to interactions and nonlinearities) for placing a *Y* or multiple *Y*s onto a specific target. They can also be used to optimally reduce variability of a critical functional response in the presence of noise factors but only on a limited number of *x* parameters and noise factors.

2.6 Validate Phase

The purpose of the Validate phase is to validate the process, assess the performance, failure modes, and risks, and iterate through a revised process until you are ready to finalize and stabilize the new process. The main activities performed in the Validate phase are

1. Validate process or product.
2. Assess performance, failure modes, and risks.
3. Iterate design and finalize.

The activities and related tools and deliverables of each activity are shown in Figure 2.29 for the Validate phase.

	Validate Activities	Tools/Deliverables
13	• Validate process	• Design of experiments (DOE) • Pilot • Statistical analysis
14	• Assess performance, failure modes, and risks	• Mistake proofing • Dashboards • Scorecards • Hypothesis tests • Analysis of variance (ANOVA)
15	• Iterate design and finalize	• Replication opportunities • Statistical process control (SPC)

FIGURE 2.29
Validate Phase Activities and Tools/Deliverables.

2.6.1 Validate Process or Product

The first activity in the Validate phase is to validate that the process or product is meeting the CTS metric targets. Developing a dashboard or scorecard to display the key metrics to management is helpful to ensure the process is performing to expectations and specifications.

The process should be piloted for some time to assess the performance of the process, or the product should be prototyped and tested. The appropriate statistical or analysis of variance (ANOVA) tests can be performed to assess the performance.

2.6.2 Assess Performance, Failure Modes, and Risks

2.6.2.1 Hypothesis Testing

The purpose of hypothesis testing is to

- Determine whether claims on process parameters are valid.
- Understand the variables of interest, the CTS characteristics.

The hypothesis test begins with some theory, claim, or assertion about a particular characteristic (CTS) of one or more populations or levels of the x (independent variable).

The null hypothesis is designated as H_0 (pronounced "H-O"), and defined as there is no difference between a parameter and a specific value. The alternative hypothesis (H_1) is designated as there is a difference between a parameter and a specific value. The null hypothesis is assumed to be true, unless proven otherwise. If you fail to reject the null hypothesis, it is not proof that the null is true.

In hypothesis testing there are two types of errors. Type I error, or alpha (α) risk, is the risk of rejecting the null hypothesis when you should not. The probability of a Type I error is referred to as alpha. Type II error or beta (β) risk is the risk of not rejecting the null hypothesis when you should. The probability of a Type II error is referred to as beta.

When performing a hypothesis test, you select a level of significance. The level of significance is the probability of committing a Type I error, or alpha, and is typically a value of 0.05 or 0.01. Figure 2.30 shows the Type I and Type II errors.

The steps for performing a hypothesis test are

1. Formulate the null and alternative hypotheses.
2. Choose the level of significance, alpha, and the sample size, n.
3. Determine the test statistic.
4. Collect the data and compute the sample value of the test statistic.
5. Run the hypothesis test in Minitab or a statistical package.

		Conclusion Drawn	
		Do Not Reject H_0	**Reject H_0**
Actual or True State	H_0 True	Correct conclusion	Type I error
	H_0 False	Type II error	Correct conclusion

FIGURE 2.30
Type I and Type II errors.

6. Make the decision. If the *p*-value is less than our significance level, alpha, reject the null hypothesis; if not, then there are no data to support rejecting the null hypothesis. Remember, if *p* is low, H_0 must go!

Some of the most common hypothesis tests are summarized in Figure 2.31.

2.6.2.2 ANOVA (Analysis of Variance)

Analysis of variance is another hypothesis test, when you are testing more than two variables of populations. Following are the steps for running ANOVA:

1. Formulate the null and alternative hypotheses.
2. Choose the level of significance, alpha, and the sample size, *n*.
3. Collect data.
4. Check for normality of data, using a test for normality.

Test Statistics	Number of Variables	Test	Parameters
Mean	1	1 sample Z	Variance
Mean	1	1 sample *t*	Variance unknown
Mean	2	2 sample *t*	Variance unknown, assume equal variances
Mean	2	2 sample *t*	Variance unknown, do not assume equal variances
Mean	2	Paired *t*-test	Paired by subject (before and after)
Proportion	1	1 Proportion	
Proportion	2	2 Proportion	
Variance	1	1 Variance (chi-square)	
Variance	2	Variance (*F*-test)	

FIGURE 2.31
Summary of hypothesis tests.

5. Check for equal variances, using an *F*-test.
6. Run the ANOVA in Minitab or another statistical package.
7. Make the decision.

Common types of ANOVA are as follows:

- *One-way ANOVA*: Testing one variable at different levels, such as testing average grade point average for high school students for different ethnicities
- *Two-way ANOVA*: Testing two variables at different levels, such as testing the average grade point average for high school students for different ethnicities and by grade level

2.6.2.3 Customer Survey Analysis

Most surveys are attribute or qualitative data, where you are asking the respondent to answer questions using some type of Likert scale, asking importance, the level of agreement, or perhaps, level of excellence.
  Ways to analyze survey data:

1. Summarize percent or frequency of responses in each rating category using tables, histograms, or Pareto charts.
2. Perform attribute hypothesis testing using chi-square analysis.

Unlike hypothesis testing with variable data, statistically with attribute data we are testing for dependence, not a difference, but you can think "makes a difference."
  We formulate our hypotheses as

- H_o: "{factor A} is independent of {factor B}"
- H_a: "{factor A} is dependent on {factor B}"

In addition to the *p* value, we use contingency tables to help understand where the dependencies (differences) exist.
  The customer survey analysis steps include

1. State the practical problem.
2. Formulate the hypotheses.
3. Enter your data in Minitab or other statistical package.
4. Run the chi-square test.
5. Translate the statistical conclusion into practical terms.

If p, the significance level, is low then reject H_o (if p is low, H_o must go). If you fail to reject the null hypothesis, H_o, that means that you fail to reject the hypothesis that the values are independent. If you reject H_o that means that they are dependent, or dependencies or differences exist.

If the process or product is not meeting expectations, further mistake proofing can be applied to reduce errors and to maintain consistency. Mistake proofing focuses on raising awareness, vigilance, and the ability to prevent errors from occurring.

When using the statistical tests, care must be taken to check if the data follow a normal distribution, and if they do not, to use the appropriate nonnormal statistical test.

2.6.3 Iterate Design and Finalize

Replication opportunities also should be assessed to determine if the same process or similar concepts can be applied to other products or processes elsewhere in the organization. Future plans for further improving the process or designing new products should also be developed.

2.7 Chapter Summary

This chapter provided a Design for Six Sigma roadmap, describing the activities and some of the tools that can be applied in the IDDOV methodology to develop a new product or design a new process.

References

Chowdhury, Subir, *Design for Six Sigma: the Revolutionary Process for Achieving Extraordinary Profits*. Chicago, IL: Dearborn Trade, 2002.

Cudney, E., Elrod, C., and Uppalanchi, A., Analyzing Customer Requirements for the American Society of Engineering Management Using Quality Function Deployment, *Engineering Management Journal*, 24(1), 2012.

Deming, W.E., *Out of the Crisis*, MIT Press, Cambridge, MA, 2000.

Evans, J., and Lindsay, M., *The Management and Control of Quality*, 5th ed. Mason, OH: South-Western Thomson Learning, 2002.

George, Michael L., Rowlands, David, Price, Mark, and Maxey, John, *Lean Six Sigma Pocket Toolbook*. New York: McGraw-Hill, 2005.

Malone, L., Class Notes, Guest Lecture, ESI 5227, University of Central Florida, Department of Industrial Engineering and Management System, 2005.

Paryani, K., Masoudi, A., and Cudney, E., QFD Application in Hospitality Industry—A Hotel Case Study, *Quality Management Journal*, 17(1), 7–28, 2010.

Pyzdek, Thomas, *The Six Sigma Handbook: A Complete Guide for Green Belts, BlackBelts, and Managers at All Levels*. New York: McGraw-Hill, 2003.

Scholtes, Peter R., Joiner, Brian L., and Streibel, Barbara J., *The Team Handbook*. Madison, WI: Oriel, 2003.

Bibiliography

Certified Six Sigma Primer, Quality Council of Indiana, 2001.

Section I

Product Design for Six Sigma Projects

3

Design of a Walker—A Design for Six Sigma Case Study

Anas AlGhamri, Çiğdem Sheffield, Lisette Sandoval, Elizabeth Cudney, and Sandra Furterer

CONTENTS

3.1 Project Overview

Current medical walkers present some features and mechanisms that demand repetitive force and motions from the user. This situation can cause discomfort and pain in the joint articulations to those who need to use mobility aids.

An ergonomic study undertaken by Hansen and Kennedy (1984) suggests that posture, force, and repetition are the main factors that influence the human body and it further is a cause of a phenomenon called cumulative trauma disorders (CTDs). In addition, the study of Leung and Yeh (2007) confirmed that CTDs are closely related to the weight of the walker and the holding gesture, hence a main cause of complaints about pain in the wrists, shoulders, back, and waist of those using walkers. These facts demonstrate the imperative need to design an ergonomic walker, which is the main objective of this project.

This project makes use of the best practices and tools from the Design for Six Sigma methodology. The gathering of needs from customers, as well as principal concerns in the use of this mobility aid were collected through a survey conducted of a representative group of 15 elderly people in Rolla, Missouri. Furthermore, the identification of technical requirements was analyzed through the quality function deployment (QFD) matrix. A posterior selection study was performed to define concepts through the use of the Pugh Concept Selection matrix. After obtaining the superior concept, a design failure mode and effects analysis (DFMEA) was performed to analyze potential failure modes presented in the design.

The project goal was to redesign walkers for elderly individuals by improving features identified by the user's survey, including weight, stability, multifunctional use, carrying function, and identification function.

A benchmark study of the current walkers was required to understand the advantages and disadvantages. The redesign of the walker was expected to combine the positive features and minimize the negative features of current walkers available in the market.

The project was limited to the redesign of a walker according to the needs expressed in the survey of a representative group of 15 elderly people in

Rolla. The project was limited to the design and construction of a prototype. No design control or transition plans for massive production are expected in the project.

3.2 Identify Phase

3.2.1 Identify Phase Activities

It is recommended that students work in project teams of three to four students throughout the Design for Six Sigma (DFSS) case study.

1. *Develop Project Charter*: Use the information provided in the Project Overview section to develop a project charter for the DFSS project.
2. *Team Ground Rules and Roles*: Develop the project team's ground rules and team members' roles.
3. *Develop Project Plan*: Develop your team's project plan for the DFSS project.

3.2.2 Identify

3.2.2.1 Project Charter

The first step was to develop a project charter.

Project Name: Redesign of walker

Project Overview: The purpose of this project is to redesign a more ergonomic walker for climbing slopes and going up and down stairs. The new design seeks as well to improve the grip support, the adjustment of height and width, and the fold-up mechanism that will be foreseen according to user specifications.

Problem Statement: Current walkers in the market present some features and mechanisms that demand force and repetitive motions from the user. This can cause discomfort and pain in the joint articulations. Hence, the project is aimed at developing an alternative design that is more ergonomic by taking into account posture, force, and repetitive motions from users.

Customer/Stakeholders: The new design is intended to be useful for people who have problems of mobility, need a walking support, and/or are undergoing rehabilitation.

Goal of the Project: The goal of the project is to redesign a walker that can suit the mobility and safety needs of the customers and at the same time have a more ergonomic and comfortable design. The project also must take into account the material, resistance, and cost of the new walker. In addition, it is intended to reduce the likelihood of accidents due to the design that may cause falls when using the walkers.

Scope Statement: The project is limited to the redesign of a walker according to customer specifications. The construction or manufacturing of the walker will not be included in the scope; however, the development of a walker prototype for the new design will be conducted.

Projected Financial Benefit(s): The more comfortable walkers available in the market are usually very expensive for seniors who usually depend on a spare pension amount or a limited income. The project team seeks to develop a new walker that can be more affordable yet ergonomic at the same time.

3.2.2.2 Team Ground Rules and Roles

The team informally developed several ground rules for the project:

- Everyone is responsible for the success of the project.
- Listen to everyone's ideas.
- Treat everyone with respect.
- Contribute fully and actively participate.
- Be on time and prepared for meetings.
- Make decisions by consensus.
- Keep an open mind and appreciate other points of view.
- Communicate openly.
- Share your knowledge, experience, and time.
- Identify a back-up resource to complete tasks when not available.

3.2.2.3 Project Plan

The design roadmap was created in phases by choosing the best practices and tools from Design for Six Sigma to achieve the redesign of the medical walker and accomplish a prototype by the end of the project. Though the project only covers the redesign and prototype, the validate phase was incorporated into the study to guarantee customer satisfaction (a verification survey) and a guide was formulated to achieve system capability (balanced scorecard).

The project plan was created by the team. The project was divided into phases and activities by using a Gantt chart. The tasks were divided among the team members, and a planning schedule was made for the completion of the different phases of the project over the weeks as shown in Figure 3.1.

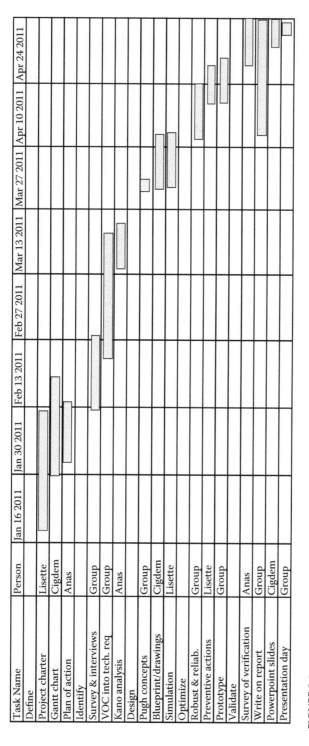

FIGURE 3.1
Project Gantt chart.

3.2.3 Identify Phase Case Discussion

1. DFSS Project Charter: Review the project charter presented.

   a. A problem statement should include a view of what is going on in the business and when it is occurring. The project statement should provide data to quantify the problem. Does the problem statement provide a clear picture of the business problem? Rewrite the problem statement to improve it.

   b. The goal statement should describe the project team's objective and be quantifiable, if possible. Rewrite the goal statement to improve it.

   c. Did your project charter's scope differ from the example provided? How did you assess what was a reasonable scope for your project?

2. Project Plan

   a. Discuss how your team developed their project plan and how they assigned resources to the tasks. How did the team determine estimated durations for the work activities?

3.3 Define Phase

3.3.1 Define Phase Activities

1. *Collect voice of the customer (VOC)*: Create a VOC survey to understand the current and potential customers' requirements.

2. *Identify critical to satisfaction (CTS) measures and targets*: Based on the VOC, determine the CTS measures and then develop targets using benchmarking data.

3. *Translate VOC into technical requirements*: Using the CTS measures and targets, identify the technical requirements for the product.

3.3.2 Define

In this first phase, the voice of the customer (VOC) was collected through a survey conducted of a representative group of 15 elderly people living in nursing homes, assisted living houses, and private houses. The survey was divided into four parts consisting of nine demographic questions, eight walker design questions, three feedback questions, and three ergonomic-related questions. Most of the questions were prepared in a multiple choice format. There were also open questions to be filled out. The demographic data of the overall group are as follows: age range: 64 to 92 years old (mean 81.7 years old), 11 females/4 males, 115 to 260 lbs (mean 187.5 lbs), 8 people own rollators, 7 people own walkers, the duration of using a walker varies between 3 weeks to more than 10 years. The usage of a walker is "very seldom" to "always used." In addition, nine people still may need to use a cane as well as their walker.

> *Checklist*: Market segment analysis was performed by visiting local medical stores and online Web sites and performing a market trend forecast to address the question, "Is there something like this out on the market yet?" In addition, benchmarking, gathering voice of the customer through survey, quality function deployment, and Kano diagrams were utilized to ensure the team was heading in the right direction.
>
> *Scorecard requirements*:

- Studying the market forecast
- Benchmarking our product to others available in the market
- Gathering information on the voice of the customer through an online survey
- Translating customer needs to useful metrics and ranking them through building our House of Quality
- Prioritizing customer requirements based on survey results
- Performing a Kano analysis
- Completing the House of Quality

3.3.2.1 Identifying "Voice of the Customer"

In order to identify the needs of the customer, the team created a survey designed to help us understand the most important factors to a walker. The survey included 17 questions as shown below.

> We are conducting a project to redesign a more ergonomic and affordable medical walker for people with mobility problems and in rehabilitation. The information provided will be used to identify satisfaction of customers with current walkers, problems in the design that can be improved, and helpful features that should be incorporated. The team will make use of Design for Six Sigma tools to translate your inputs into design requirements in order to create a more ergonomic and safe walker yet at a low cost with innovative features. We appreciate your time and willingness to complete the following survey. The survey is divided into four parts with a total of 17 questions. It should not take more than 20 minutes.

I. Demographic Questions
   1. What is your age group?
      a. Less than 65
      b. 66 to 75
      c. 76 to 85
      d. Over 85
   2. What is your weight?
      a. Under 140
      b. 141 to 165
      c. 166 to 190
      d. Over 190
   3. What kind of a walker do you use?
      a. Walker with wheeled legs
      b. Walker without wheeled legs
      c. Rollator
   4. How long have you been using a walker?
      a. Less than 1 year
      b. 1 to 3 years
      c. 3 to 5 years
      d. Over 5 years
   5. How often do you use your walker?
      a. Always
      b. Only indoors
      c. Only outdoors
      d. When I feel weak
   6. Have you suffered an injury while using a walker? (Yes/No)
      If yes, please explain.

II. Walker Design Questions
   1. Is your walker easy to open and fold up?
      a. Easy
      b. Somewhat easy
      c. Difficult
      d. Very difficult
      Why?
   2. Is your walker easy to adjust, for example, could you adjust the height and width of your walker without assistance?
   3. What would you change on the grip support? What do you like about it? Has your grip support worn out with time, making it slippery in the support?
   4. What do you think about having these features in your walker (*choose three*):
      – Wheels and brakes
      – Different color
      – Lights
      – Adjustable width
      – Stress relief material in the grip support
      – Hook to hang personal and/or medical items
      – Basket in front of the walker
      – Special kind of legs that do not scratch the floor

5. Pros of your walker (please select all that apply)
   a. Light in weight
   b. Solid and robust design
   c. Stores easily
   d. Easy to use
   e. Safe
   f. Supports weight
6. Cons of your walker (please select all that apply)
   a. None
   b. Difficult to store
   c. Difficult to use
   d. Unsafe
   e. Hard to drag
   f. Leaves marks on my floor
   g. It rattles
   h. Other (Please specify)

III. Feedback Questions
1. What are the most annoying things about walking with your walker indoors and outdoors?
2. What features would you change in your walker? Why?

IV. Ergonomic Study Questions
1. Is your walker (or rollator) hard to lift as you walk?
2. What repetitive movements does your walker make you do?
   a. Holding gesture (holding in the grip support)
   b. Braking the walker (only applicable for rollators with braking system of hand)
   c. Other (Please specify)
3. How stable is your walker when walking on the ground (flat terrain)?
   a. Very stable
   b. Stable
   c. Not very stable
   d. Not stable at all

3.3.2.2 Quality Function Deployment

Using the voice of the customer we built the House of Quality. These are the steps we used to build it:

Step 1: List customer requirements.

Step 2: List technical descriptors (characteristics that will affect more than one of the customer requirements, in development or production).

Step 3: Compare the two (customer requirements to technical descriptors) and determine relationships.

Step 4: Develop the positive and negative interrelated attributes and identify "trade-offs."

Customer needs are analyzed through the use of QFD to identify functional requirements and translate them into technical requirements for the design. A snapshot of the House of Quality can be seen in Figure 3.2.

3.3.2.3 Product Design Metrics and Kano Analysis

The next step was to develop the characteristics into metrics to identify the ideal specifications to meet customer requirements. For the design requirements, the team used the Kano analysis as shown in Figure 3.3.

3.3.3 Define Phase Case Discussion

1. How did your team perform the VOC collection? How could VOC collection be improved?
2. Did your team create and distribute a customer survey, and if so, what is the appropriate statistical analysis to perform to identify the importance of the customers' requirements?
3. Did you perform a QFD? How did you identify the technical requirements, and the correlations between customer and technical requirements?
4. What is the value of using the Kano model in your VOC analysis?

3.4 Design Phase

3.4.1 Design Phase Activities

1. Identify process elements.
2. Design process.
3. Identify potential risks and inefficiencies.

3.4.2 Design

The Design stage of the project assures the subsystem concept and design development of the walker. First, the team conducted subsystem benchmarking in order to develop important and critical functions that align with customer requirements.

> *Checklist*: Concept generation technique (TRIZ, brainstorming), Design for Manufacture and Assembly, concept generation, affinity diagram, Pugh concept evaluation and selection.

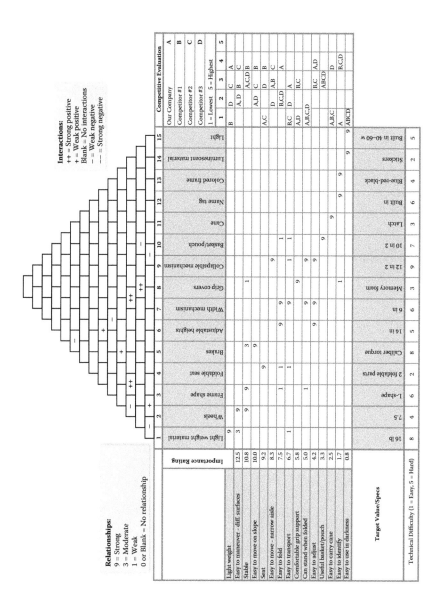

FIGURE 3.2
House of Quality.

Customer Requirements	New Needs	Unique Needs	Difficult Needs	Noncritical Needs
Light weight			X	
Easy to maneuver on different surfaces			X	
Maintains stability			X	
Easy to move on slopes	X		X	
Provides a seat		X		
Easy to move through narrow aisles			X	
Easy to fold			X	
Easy to transport			X	
Provides a comfortable grip support	X		X	
Stands when folded	X		X	
Easy to adjust height			X	
Provides a useful basket/pouch				X
Easy to carry cane			X	X
Easy to identify	X			X
Easy to use in darkness	X	X		X

FIGURE 3.3
Kano analysis.

Scorecard requirements:

- Generating seven design concepts that meet the customer requirements
- Evaluating superior concepts and superior technology to beat the market
- Analyzing and studying superior concepts—looking at the competitors
- Adding value to the design by thinking "outside the box"

3.4.2.1 Subsystem Benchmarking Analysis

The following subsystems are critical for customer requirements:

- *Hand grip*: Grip support should be an ergonomic and compliant material. This feature was inspired from walking poles.
- *Back support*: Flexible fabric provides back support. It will permit folding and reduce the walker weight.

- *Seat*: Cloth seats could make the customers feel unsafe. The design of the walker should include two firm padded seat cushions in order to divide the two parts for easier folding which would eliminate customer complaints regarding an unsafe feeling with cloth seats.
- *Collapsing mechanism*: The user should be able to collapse the walker without excessive force and without having difficulty. A simple strap will provide a lightweight and robust design.
- *Horizontal support bar*: A plain scissor jack mechanism will provide safe and constant separation distance.
- *Front wheel assembly*: The wheels should be large wheels with a full swivel feature and high-profile pneumatic tires.
- *Rear wheel assembly*: The wheels should be large wheels with a nons-wivel feature and high-profile pneumatic tires.
- *Brake system assembly*: Metal contact pads for pneumatic tires with a trailing edge feature to provide quick stop in reverse motion and smooth stop in forward motion. The brakes should have a locking mechanism, for example, while the user is using the seat.
- *Frame*: Tubular aluminum similar to a bike frame created from a basic truss.

3.4.2.2 Concept Generation

The goal of our project was to redesign a walker to meet customer requirements best. Keeping this in mind, the team brainstormed a few design concepts that might function better than the current walkers. All the design concepts were generated by dividing the whole system into nine subsystems as defined in the previous section. The team used the Pugh Concept Selection matrix to analyze candidate design concepts based on the VOC requirements to get a superior concept for the design.

3.4.2.3 Development of Functions

The subsystem benchmarking analysis allows identification of critical requirements to achieve customer requirements. The functions were created after a long brainstorming exercise by the team taking into account criticality and independence from the current concepts available in the market as shown in Figure 3.4.

3.4.2.4 Design Concepts

Concept A: This will be four-wheeled with a fabric seat (for convenient folding) and locking handbrakes for any use (indoors and outdoors). Details: The handgrip features will be compliant, ergonomic grips,

Subsystem	Function	Voice of the Customer (VOC)
Hand grip support	Height, material, design	Easy to maneuver
Back support	Height, position, inclination degree, and material of assembly	Comfortable and safe back support
Seat assembly	Height, width, material, and stability	Stable seat
Elbow rest	Height, design, stability	Lower grip support for rising from chairs
Bar mechanism for folding	Length bars, angle of amplitude	Easy to fold
Front wheel assembly	Stability, wheel size	Easy to turn
Rear wheel assembly	Speed, wheel size, and direction of degree of freedom	Easy to move
Brake system assembly	Pressure brake according to customer strength	Easy to brake
Frame	Material, stability, design, angle	Lightweight, easy to transport, easy to move through narrow aisle

FIGURE 3.4
Product functions.

with a durable hard inner core. The handlebar is swept back toward the user, with each grip ending 30° from parallel to each other. The back support has a soft flexible strap to support the user's back while sitting. The seat will be made out of plastic that can be folded, while collapsing the folder. The collapsing mechanism will be easy for the user to collapse. A simple strap on the seat will enable the walker to collapse. The design will include 4-inch swivel-wheels with high-profile pneumatic tires for the front assembly and 4-inch nonswivel wheels with high-profile pneumatic tires for the rear assembly. Metal contact pads are for pneumatic tires with a trailing edge feature to provide quick stops in reverse motion and smooth stops in forward motion. The brakes should have a locking mechanism, for example, while the user is using the seat. A tubular aluminum frame is used. A mesh pouch will be placed in front of the seat. Concept A is shown in Figure 3.5.

Concept B: This will be four-wheeled with a seat, back support, and locking handbrakes for any use. Details: The handgrip features will be compliant, ergonomic grips with a durable hard inner core. The handlebar is swept back toward the user, with each grip ending 30° from parallel to each other. The back support has a soft flexible strap to support the user's back while sitting. The seat will be two

FIGURE 3.5
Concept A.

firm pads that are placed on top of a fabric. The collapsing mechanism will be easy for the user to collapse. A simple strap on the seat will enable the walker to collapse. The horizontal support bar will be a plain scissor jack mechanism. The front and rear wheel assemblies will have 4-inch swivel-wheels with high-profile pneumatic tires. The frame will be tubular aluminum, and the shape is inspired from a crutch and walking poles. The brake assembly will be metal contact pads for pneumatic tires with a trailing edge feature to provide quick stops in reverse motion and smooth stops in forward motion. The brakes should have a locking mechanism, for example, while the user is using the seat. A mesh pouch will be placed in front of the walker. The walker will have an elbow rest assembly to give support to the user while indoors. The elbow rest assembly will rotate 180° horizontally and vertically so that the user can use the walker outdoors as well. The walker will have built-in light-emitting diode (LED) lights placed on both vertical bars, as well as light reflectors on the lower bars of the walker. The user will be able to attach his or her cane, trigger, and other medical assistive tools on the walker. It can be adjusted for the user's height. Concept B is shown in Figure 3.6.

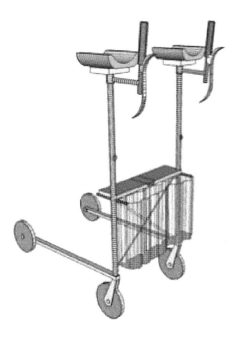

FIGURE 3.6
Concept B.

Concept C: Concept C will be four-wheeled with a seat and locking handbrakes. Details: The handgrip features will be compliant, ergonomic grips with a durable hard inner core. The handlebar is swept back toward the user, with each grip ending 30° from parallel to each other. The back support has a soft flexible strap to support the user's back while sitting. The seat will be two firm pads that are placed on top of a fabric. The collapsing mechanism will be easy for the user to collapse. A simple strap on the seat will enable the walker to collapse. The horizontal support bar will have a hinge that will permit collapsing. The front wheel assemblies contain 4-inch swivel-wheels with high-profile pneumatic tires. The rear wheel assemblies contain 4-inch nonswivel wheels with high-profile pneumatic tires. The frame is tubular aluminum. The brake cables will be hidden inside the frame tubes. The cane and trigger holder tool will be built in the frame tubes. There will be a mesh pouch under the seat. The walker will be fully collapsible. There will be a snap to keep the walker collapsed to protect the user while lifting. The walker will be available in a variety of colors. Concept C is shown in Figure 3.7.

Concept D: Concept D will be two-wheeled with a seat, pressure brakes, and a lower set of handles for assisting support when rising in seat or toilet (especially for indoor use). Details: The handgrip features will be compliant, ergonomic grips with a durable hard inner core.

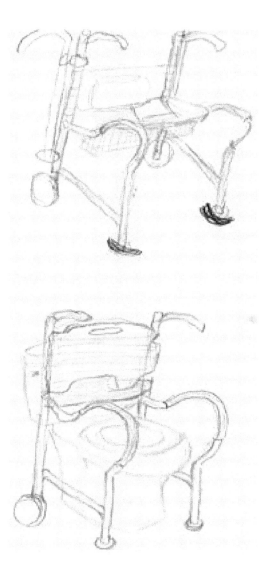

FIGURE 3.7
Concept C.

The handlebar is swept back toward the user, with each grip ending 30° from parallel to each other. The back support has a molded back support. The seat will be one piece made out of sturdy plastic. The folding mechanism will be a latch on both side bars, which will enable the user to fold the walker. The seat and the molded back support will hold the side bars, which will enable the walker to stay stable. Four-inch swivel-wheels with high-profile pneumatic tires will be in the front assemblies, and there will not be rear wheels.

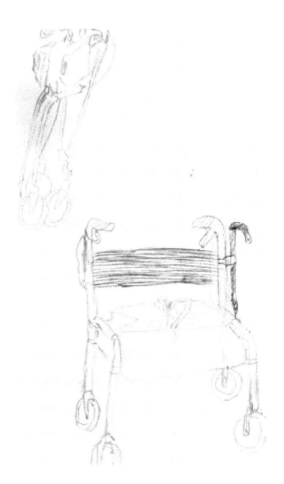

FIGURE 3.8
Concept D.

Instead, the rear legs will have glides attached to allow the user to move on different surfaces smoothly. It will have a tubular aluminum frame. The walker will have a lower set of handles to assist the user in rising from seats using less arm strength. The brake cables will be hidden inside the frame. There will be a pouch under the seat. The molded back support can easily be turned into a tray. The medical assistive tools can be attached to it. Concept D is shown in Figure 3.8.

Concept E: Concept E will have bigger front wheels and swivel rear wheels with a seat specifically designed for outdoor use. Details: There will be a one-piece handle bar similar to a shopping cart. The back support has a soft flexible strap to support the user's back while sitting. The seat will consist of three pieces and be firm plastic. The user will lift

FIGURE 3.9
Concept E.

the seat up, and the middle part of the seat will enable the collapsing. The top bar, the seat, and the X-bar mechanism will keep the walker stable and open. The front wheel assembly contains 3-inch swivel-wheels with high-profile pneumatic tires. The rear wheel assembly contains 4-inch nonswivel wheels with high-profile pneumatic tires. The brake cables will be hidden inside the frame tubes. The cane and trigger holder tool will be built in the frame tubes. There will be a snap to keep the walker collapsed to protect the user while lifting it in and out of the vehicle. The walker will have a variety of colors. Concept E is shown in Figure 3.9.

Concept F: Concept F will be three-points with no seat for convenient maneuverability through narrow spaces indoors and/or outdoors. The handgrip features will be compliant, ergonomic grips with a durable hard inner core. The handlebar is swept back toward the user, with each grip ending 30° from parallel to each other. The user will lift the latch up, which is located on the horizontal bar. There are 4-inch swivel-wheels with high-profile pneumatic

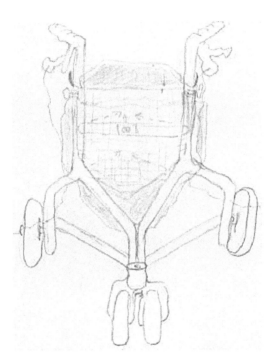

FIGURE 3.10
Concept F.

tires for both the front and rear wheels. The front wheels will not
be aligned with the rear wheels to give more maneuverability to
the user (inspired from the shopping cart). The cane and trigger
holder tool will be built in the frame tubes. There will be a snap
to keep the walker collapsed to protect the user while lifting. The
walker will have a variety of colors. The walker will have a mesh
pouch, which will be folded during collapsing. Concept F is shown
in Figure 3.10.

Concept G: Concept G will be two-wheeled with no seat for easy collapsing and convenient traveling. The handgrip features will be compliant, ergonomic grips with a durable hard inner core. The handlebar
is swept back toward the user, with each grip ending 30° from parallel to each other. The user will lift the latch up which is located on
the horizontal bar. There are 4-inch swivel-wheels with high-profile
pneumatic tires for the front assembly, and the rear legs will have
glides to give a smooth ride to the user on different surfaces. The
cane and trigger holder tool will be built in the frame tubes. There
will be a snap to keep the walker collapsed to protect the user while
lifting. The walker will have a variety of colors. The walker will

FIGURE 3.11
Concept G.

have a mesh pouch on each side of the walker, which will be folded during collapsing. The walker is inspired by an umbrella stroller. Concept G is shown in Figure 3.11.

3.4.2.5 Pugh's Concept Selection

Through our product development the team created seven different designs of the walker. To determine the concept that had the greatest potential for success, the team utilized the Pugh Concept Selection matrix to assist in narrowing the scope to a single design as shown in Figure 3.12.

The initial best-in-class benchmarked datum concept was selected based on the preference of the surveyed participants for four-wheeled walkers or rollators with brakes and seats. Eight participants owned a rollator and expressed satisfaction, comfort, and practical and easy use. In addition to the benchmark study of walkers in the market, the four-wheeled or rollator was also confirmed to be the best mobility aid among walkers; hence, it was chosen as the datum.

Concept B was selected as the best among other designs because it exceeded the features of the best-in-class rollator. From the sum of (+) and (–), the design team was able to identify weaknesses in low-scoring concepts that can be turned into (+) such as the ability to move in different grounds and through narrow aisles, and it definitely requires wheels, seat, and extra support. The team also identified weaknesses in high-scoring concepts that

Criteria	Design Concepts							Datum
	A	B	C	D	E	F	G	
Light weight	–	–	S	–	S	+	+	
Easy to maneuver on diff. surfaces	S	S	–	S	S	–	–	
Stable	S	S	+	S	S	–	–	
Easy to move on slope	S	S	–	S	S	–	–	
Seat	–	–	–	S	–	–	–	
Easy to move through narrow aisle	S	S	S	S	S	+	–	
Easy to fold	S	S	S	+	+	S	+	
Easy to transport	+	+	S	+	–	+	+	
Comfortable grip support	S	+	S	S	–	S	–	
Can stand when folded	+	+	–	–	+	+	–	
Easy to adjust	S	S	S	S	S	S	S	
Useful basket/pouch	+	+	S	+	–	+	–	
Easy to carry cane	S	S	+	S	S	S	S	
Easy to identify	+	+	–	–	S	S	–	
Easy to use in darkness	+	+	S	S	S	S	S	
Sum of (+)	5	6	2	3	2	4	3	
Sum of (–)	2	2	5	3	4	4	9	
Sum of (S)	8	7	8	9	9	6	3	

FIGURE 3.12
Pugh's Concept Selection matrix.

can be turned into (+), such as improving the weight of the walker and incorporating a comfortable seat.

After this analysis, a hybrid super concept was created by choosing most of the features of Concept B with additional strength features from other concepts. According to the design of Concept B, some features are more likely to fit, improve, and balance a realistic integration with such a concept. This is the reason that the design team chose Concept B as the super concept with the following extra features to eliminate the weaknesses in weight and seat: aluminum material with limited bars and joint assembly in the frame and comfortable seat.

3.4.2.6 Detailed Model of the Product Design

Once the components were identified, the team developed a detailed model for the required design specifications as shown in Figure 3.13.

3.4.3 Design Phase Case Discussion

1. How did you generate your design concepts?
2. How did you determine how your concepts compared using the Pugh Concept Selection matrix?
3. How did you derive the best combination of your design elements from each concept?

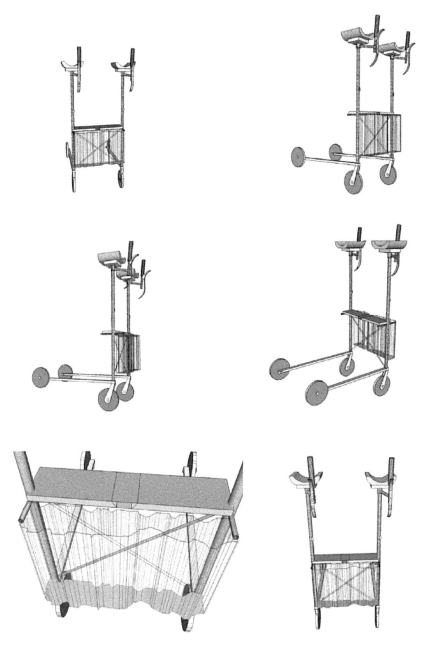

FIGURE 3.13
Final concept model.

3.5 Optimize Phase

3.5.1 Optimize Phase Activities

1. Implement pilot process.
2. Assess process capabilities.
3. Optimize design.

3.5.2 Optimize

In this phase, the team reviewed the selected superior concept to identify possible flaws and their effects and probable solutions on how to reduce them. The team also compared the influence of one factor on another and outside noise factors.

> Checklist: Design of experiments, failure mode effect analysis, analysis of mean, analysis of variance, design capability studies, critical parameter management
>
> *Scorecard requirements*:

- Identifying noise factors and control factors from both customers and competitors
- Lowering occurrences of high-severity, high-occurrence defects
- Identifying factors that influence each other
- Determining probable solutions to failure

3.5.2.1 Design Failure Modes and Effects Analysis

The design failure mode and effects analysis (DFMEA) was performed in order to quantify the reliability of the main components/parts of the walker. The DFMEA scores for severity, occurrence, and detection categories were applied for the scale 1 to 5, which counts 1 as good and 5 as very bad. The severity scores that are greater than three are taken as critical because the severity of a failure can potentially harm the user. The DFMEA is shown in Figure 3.14.

All high-priority issues identified in the DFMEA are those failure modes that have a severity rating equal or greater than four. Figure 3.15 contains a detailed plan to reduce the risk priority number (RPN) values and make the walker design more insensitive to noise without removing the sources of variation.

The priority identification and highest RPN values allow for identification of the critical subsystems of the walker design. First, in the particular design that the team is designing, the frame and collapsing mechanism are the main subsystems bearing most of the weight load due to sitting and

FAILURE MODE AND EFFECT ANALYSIS

___Medical Walker_____

Design Responsibility___Walker group_____

Key Date_ 04/26/2011_____

Core Team:___Team No. 3

Part or Component Identification	Function	Potential Failure Mode	Potential Effect(s) of Failure	SEV	Potential Cause(s)/Mechanisms of Failure	OCC	Current Detection Controls	DET	RPN	Preventive Recommended Actions
Hand grip	Provides support and guidance to conduct the walker	Loose	Lack of support and risk of injury in user's wrist	4	Severe temperature change (hot softens, cold hardens)	2	Non-circular cross section	2	16	Insure hand grip is non-load bearing
		Deformation or cracking	Discomfort and injury to the user	3	UV Hardening	2	None	5	30	UV Compliant Materials
Back support	Provides support to the back of the user	Height of the back support is in improper position	Results in a fall	5	User did not check the suitability of back support prior to sitting	2	None	5	50	Have the back support feature be adjustable according to user's height and weight.
Firm padded seat	Allows support for resting	The fabric tears or stitching fails	Results in a fall	5	Rottens or UV Hardening	2	Visual Inspection	3	30	Use Kevlar fabric under the padded seat
Elbow rest assembly	Provides support to elbows when seated and support to hands and body	Loose	Lack of support and risk of injury in user's wrist	4	Severe temperature change (hot softens, cold hardens)	2	Non-circular cross section	2	16	Insure hand grip is non-load bearing
		Deformation or cracking	Discomfort and injury to the user	3	UV Hardening	2	None	5	30	UV Compliant Materials
Horizontal Stabilizer Bar:	Provides horizontal stability	Scissor pin fails	Catastrophic failure	5	Fatigue	2	None	5	50	Redundant scissor
		Scissor bar buckles	Collapsing function fails	2	Damage during storage	4	Visual Inspection	3	24	Locking strap maintain collapse state
Front wheel assembly	Provides degrees of freedom to turn or move into different directions	Flat tire	Slow motion	2	Air diffuses	1	Check wheels regularly	2	4	Check wheels regularly
Rear wheel assembly	Provides degree of freedom in one single direction	Flat tire	Slow motion	2	Air diffuses	1	Check wheels regularly	2	4	Check wheels regularly
Brake system assembly	Allows user to stop or reduce mobility	Cable loosens	Reducing braking forces	3	Cable stretches	1	Feel	3	15	Adjustable tension knots
		Brake failure	Fall	5	Brakes got wet	2	Check read on tire if smooth	2	20	Use roughen metal surface on brake bar.
Walker frame	Provides stability and maneuverbility	Aluminium scratches	User gets cut and potential infection	5	Rough handling	1	None	5	25	Paint aluminum
									314	

FIGURE 3.14
Design FMEA for walker.

Subsystem	Failure Mode	Potential Effects of Failure	Current Control	Risk Priority Number (RPN)	Recommended Actions
Back support	Height of the back support is in improper position	Results in a fall	None	50	Have the back support feature be adjustable according to user's height and weight
Horizontal stabilizer bar	Scissor pin fails	Catastrophic failure	None	50	Redundant scissor
Handgrip	Deformation or cracking	Discomfort and injury to the user	None	30	Ultraviolet (UV)-compliant materials
Firm padded seat	The fabric tears or stitching fails	Results in a fall	Visual inspection	30	Use Kevlar fabric under the padded seat
Elbow rest assembly	Deformation or cracking	Discomfort and injury to the user	None	30	UV-compliant materials
Walker frame	Aluminum scratches	User gets cut and potential infection	None	24	Paint aluminum
Brake system	Brake failure	Fall	Feel	25	User rough metal surface on brake bar

FIGURE 3.15
Design improvement plan.

walking support. These two subsystem designs are critical also for other subsystems that depend on them as main structures of support such as the hard seat and back support. If the frame and horizontal support bar present a failure, the whole structure may collapse with the subsequent danger of harming the user.

Another critical mechanism is the brake system. It is highly important to verify the design through testing the product for reliability and robustness in order to guarantee safe use of a front-swivel-wheel walker. The functionality of the brake assembly should be tested in different types of grounds and slopes; the design must consider using the walker in handicapped ramps. In addition, it is critical to test the pressure and reaction time that the walker requires to slow down or stop. These functions must align with the force applied by users.

The two remaining critical subsystems are the seat and back support. These subsystems depend mainly on the stability and structure support of the frame and folding mechanism. However, the seat and back support need special height, position, and inclination angle to assure these subsystems align with the overall stability of the frame and folding mechanism.

3.5.3 Optimize Phase Case Discussion

1. How did you define a failure?
2. How did you determine which factors were significant to your design?
3. How did you determine the design solutions to prevent defects?

3.6 Validate Phase

3.6.1 Validate Phase Activities

1. Validate process.
2. Assess performance, failure modes, and risks.
3. Iterate design and finalize.

3.6.2 Validate Phase

In this phase, the team verified the final design parameters with responses from customers by providing them with a prototype.

Checklist: Measurement system analysis, manufacturing process capability study, reliability assessment, worst-case analysis, analytical tolerance design

Scorecard requirements:

- Prototype approved by customer
- Meets large portion of customer needs
- Ease to manufacture and cost estimate
- Reliability performance

3.6.2.1 Construction of Pilot Prototype

A pilot prototype was created to demonstrate the main subsystems and components of the walker design. The creation of the prototype is helpful

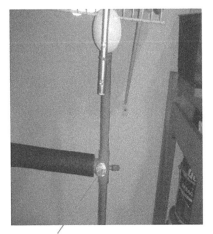
Light reflectors

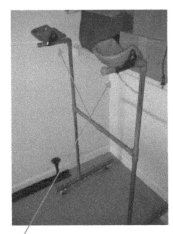
Grip supports (Knee pads)

Hand brakes (hose nozzles)

FIGURE 3.16
Prototype.

to identify components that could be combined to integrate functions, separated to improve functions, or improved in the design to facilitate the production of parts.

3.6.2.2 Materials Used for the Prototype

- The design team used two polyvinyl chloride (PVC) pipes of 1′ and 5′ diameter to create the frame.
- The grip support was created using knee pads.

- The collapsing mechanism consisted of a plastic hose of a smaller 1' diameter, which was inserted into a three-part bar in order to depict the latch to lock the walker when folding.
- The wheels were assembled to the PVC pipes. Note that though the prototype contains four swivel wheels due to trouble finding nonswivel wheels, the actual design considers the two rear wheels to be nonswivel for better control of the walker.
- The lights were actual light reflectors on each side of the frame.

Figure 3.16 shows the pilot prototype.

A survey for verification was created to validate customer requirements with the new walker design. A sample population of five users participated in the survey whose age ranged between 64 and 73 years old (mean was 67.6, standard deviation was 3.78). The questions in the survey were limited to six questions regarding what features of the new walker design people like and dislike most. The results showed that participants found more of the following characteristics appealing: the built-in light features (lighting system) and the size of the walker after collapsing which makes it more convenient for transport. However, the main concerns that participants have expressed in the new design are the flexible back strap and pouch features. According to the surveyed customers, the flexible back strap could be unsafe and offer lack of support. One of the answers expressed by one participant: "I wouldn't lie back on a flexible strap. I want a firm rest support, not something that will make me nervous of a fall or uncomfortable altogether." In addition, one of the answers expressed about the pouch in front of the walkers: "I probably wouldn't see much if you place a pouch in front of the walker. How am I going to be sure of what is ahead of me if the pouch will block my vision?"

The main concerns have been addressed by the design team through a brainstorming session to generate alternative ideas in the design reaching the following changes in the design:

- Flexible strap
  - Replace it by hard back support
- Pouch
  - Relocate smaller pouches in the sides of the walker
- Relocate pouch under the seat

3.6.3 Validate Phase Case Discussion

1. How did you identify potential failure modes of your product?
2. How did you identify potential risks of your product?
3. How would you assess the potential market for your product?

3.7 Summary

The use of DFSS tools allowed the team to understand and interpret customer needs, as well as to translate those needs into technical design requirements. The breakdown of systems into subsystems facilitated the identification and design on individual units. Concept generation allowed the team to integrate better features in the walker or rollator. The Pugh Concept Selection matrix was useful in the selection of the superior concept. DFMEA allowed the team to address possible design failures in the superior concept. The verification survey helped the team identify parts of the design that were not satisfied by the customer and replace these with alternative designs. The goal of redesigning a walker or rollator based on features stated by the 15 participants in the questionnaire was achieved.

References

Hansen, D.J., and Kennedy, K.W., Repetitive trauma disorders: Job evaluation and design. *Human Factors*, 28(3), 325–336, 1984.

Leung, C.Y., and Yeh, P.C., A Study on the Potential Risk Factors for the Elders Using Walkers, presented at the International Design Conference, Cumulus Kyoto, 2007.

4

Design of a Military Tool Holder—A Design for Six Sigma Case Study

David Moore, Antonio Ward, Bruce Lane, Thomas Fitzpatrick, Elizabeth Cudney, and Sandra Furterer

CONTENTS

4.1 Project Overview

The purpose of this project was to redesign the existing Modular Lightweight Load-carrying Equipment (MOLLE) or Pouch Attachment Ladder System (PALS) for a military tool kit holder. The intent was to develop a quick-release system that could be used with the current MOLLE system.

The original military tool kit holder system does not allow for a quick attachment or detachment, it is a semipermanent attachment system. The tool holder has web straps (PALS) that can be attached to most MOLLE systems. This current system is not practical due to its constraints on the individual's movements and the individual's safety; it also negatively impacts the morale of the carrier. The original system does not meet the needs of the modern soldier or law enforcement officer working in a dynamic environment.

The project was limited to the military tool kit holder. The design team developed multiple new designs in order to provide a recommendation for an improved military tool kit holder attachment system. The parameters were limited to the method of attachment between the military tool kit holder and the current Outer Tactical Vest (OTV) or other PALS equipment. A new design will be developed that meets the individual's requirements of safety, efficiency, speed, and ease of use.

The team utilized the Design for Six Sigma (DFSS) process to design attachment concepts for the military tool kit holder. The target was to use quick-release detachable joints that are specifically focused on the tactical service member, which enables that person to perform his or her mission. The usage of this innovative technology will allow service members to accomplish their mission effectively and safely. By analyzing several different attachments, the team compared the practicality, durability, and functionality of the designs to present an optimal solution for the design. The team made incremental improvements and developed a final design of the optimal solution that addresses all of the customer functionality needs.

4.2 Identify Phase

4.2.1 Identify Phase Activities

It is recommended that students work in project teams of three to four students throughout the DFSS case study.

1. *Develop Project Charter*: Use the information provided in the Project Overview section to develop a project charter for the DFSS project.
2. *Team Ground Rules and Roles*: Develop the project team's ground rules and team members' roles.
3. *Develop Project Plan*: Develop your team's project plan for the DFSS project.

4.2.2 Identify

4.2.2.1 Project Charter

The first step was to develop a project charter.

Project Name: Quick-Release Military Tool Kit Holder

Project Overview: The purpose of this project is to redesign the existing MOLLE attachment system for the military tool kit holder.

Problem Statement: The original military tool kit holder system does not allow for a quick attachment or detachment; it is a semipermanent attachment system. The tool holder has web straps that can be attached to any MOLLE system. This current system is not practical due to its constraints on the individual's movements and the individual's safety, and it negatively impacts the morale of the carrier. The original system does not meet the needs of the modern soldier or law enforcement officer working in a dynamic environment.

Customer/Stakeholders: Military and law enforcement personnel

Goal of the Project: A new design will be developed that meets the individual's requirements of safety, efficiency, speed, and ease of use.

Scope Statement: The project is limited to the military tool kit holder. The Design for Six Sigma team will redesign and recommend an improved military tool kit holder. The parameters will be limited to the method of attachment.

Projected Financial Benefit(s): Increase demand for the military tool kit holder.

4.2.2.2 Team Ground Rules and Roles

The team informally developed several ground rules for the project:

- Everyone is responsible for the success of the project.
- Listen to everyone's ideas.
- Treat everyone with respect.
- Contribute fully and actively participate.
- Be on time and prepared for meetings.
- Make decisions by consensus.
- Keep an open mind and appreciate other points of view.
- Communicate openly.
- Share your knowledge, experience, and time.
- Identify a backup resource to complete tasks when not available.

4.2.2.3 Project Plan

The project was set into the DFSS phases of Identify-Define-Design-Optimize-Validate (IDDOV). Specific requirements were identified for this project and are represented in their respective phase of the DFSS project. The phases of this project are Identify, Define, Design, Optimize, and Verify. After each, a recap or gate will identify the benchmarks and issues with product design and development processes.

A Gantt chart was used to keep the project development cycle on track and ensure that all key steps in the process were accomplished as shown in Figure 4.1. This chart includes dates and tasks that are essential for appropriate usage of time management during the scheduled period of work.

4.2.3 Identify Phase Case Discussion

1. DFSS Project Charter: Review the project charter presented.
    a. A problem statement should include a view of what is going on in the business, and when it is occurring. The project statement should provide data to quantify the problem. Does the problem statement provide a clear picture of the business problem? Rewrite the problem statement to improve it.
    b. The goal statement should describe the project team's objective and be quantifiable, if possible. Rewrite the goal statement to improve it.
    c. Did your project charter's scope differ from the example provided? How did you assess what was a reasonable scope for your project?

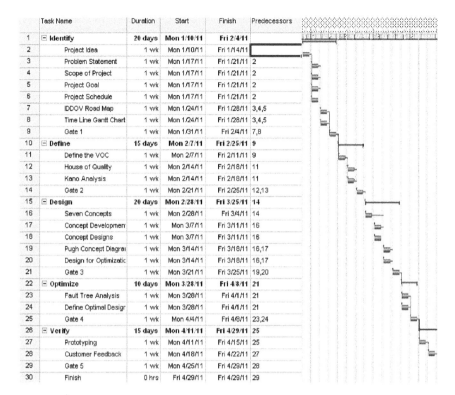

	Task Name	Duration	Start	Finish	Predecessors
1	⊟ **Identify**	**20 days**	**Mon 1/10/11**	**Fri 2/4/11**	
2	Project Idea	1 wk	Mon 1/10/11	Fri 1/14/11	
3	Problem Statement	1 wk	Mon 1/17/11	Fri 1/21/11	2
4	Scope of Project	1 wk	Mon 1/17/11	Fri 1/21/11	2
5	Project Goal	1 wk	Mon 1/17/11	Fri 1/21/11	2
6	Project Schedule	1 wk	Mon 1/17/11	Fri 1/21/11	2
7	IDDOV Road Map	1 wk	Mon 1/24/11	Fri 1/28/11	3,4,5
8	Time Line Gantt Chart	1 wk	Mon 1/24/11	Fri 1/28/11	3,4,5
9	Gate 1	1 wk	Mon 1/31/11	Fri 2/4/11	7,8
10	⊟ **Define**	**15 days**	**Mon 2/7/11**	**Fri 2/25/11**	**9**
11	Define the VOC	1 wk	Mon 2/7/11	Fri 2/11/11	9
12	House of Quality	1 wk	Mon 2/14/11	Fri 2/18/11	11
13	Kano Analysis	1 wk	Mon 2/14/11	Fri 2/18/11	11
14	Gate 2	1 wk	Mon 2/21/11	Fri 2/25/11	12,13
15	⊟ **Design**	**20 days**	**Mon 2/28/11**	**Fri 3/25/11**	**14**
16	Seven Concepts	1 wk	Mon 2/28/11	Fri 3/4/11	14
17	Concept Development	1 wk	Mon 3/7/11	Fri 3/11/11	16
18	Concept Designs	1 wk	Mon 3/7/11	Fri 3/11/11	16
19	Pugh Concept Diagram	1 wk	Mon 3/14/11	Fri 3/18/11	16,17
20	Design for Optimization	1 wk	Mon 3/14/11	Fri 3/18/11	16,17
21	Gate 3	1 wk	Mon 3/21/11	Fri 3/25/11	19,20
22	⊟ **Optimize**	**10 days**	**Mon 3/28/11**	**Fri 4/8/11**	**21**
23	Fault Tree Analysis	1 wk	Mon 3/28/11	Fri 4/1/11	21
24	Define Optimal Design	1 wk	Mon 3/28/11	Fri 4/1/11	21
25	Gate 4	1 wk	Mon 4/4/11	Fri 4/8/11	23,24
26	⊟ **Verify**	**15 days**	**Mon 4/11/11**	**Fri 4/29/11**	**25**
27	Prototyping	1 wk	Mon 4/11/11	Fri 4/15/11	25
28	Customer Feedback	1 wk	Mon 4/18/11	Fri 4/22/11	27
29	Gate 5	1 wk	Mon 4/25/11	Fri 4/29/11	28
30	Finish	0 hrs	Fri 4/29/11	Fri 4/29/11	29

FIGURE 4.1
Project Gantt chart.

2. Project Plan

  a. Discuss how your team developed their project plan and how they assigned resources to the tasks. How did the team determine estimated durations for the work activities?

4.3 Define Phase

4.3.1 Define Phase Activities

1. *Collect voice of the customer (VOC)*: Create a VOC survey to understand the current and potential customers' requirements.

2. *Identify critical to satisfaction (CTS) measures and targets*: Based on the VOC, determine the CTS measures and then develop targets using benchmarking data.

3. *Translate VOC into technical requirements*: Using the CTS measures and targets, identify the technical requirements for the product.

4.3.2 Define

The voice of the customer is extremely important when developing a product or service that provides satisfaction and innovative features. The primary focus here can be analyzed from several customer experiences, surveys, and reports. During this stage the team wanted to gather as much research from a certain population of individuals. A survey was distributed to a pool of customers and from their feedback the team was able to determine what functions the customer really wanted the military tool kit holder clip to perform.

> *Checklist*: Market segment analysis by interviewing customers and performing a market trend forecast to address the question, "Is there something like this out on the market yet?" In addition, benchmarking, gathering voice of the customer through a survey, affinity diagram, quality function deployment, and Kano diagrams were utilized to ensure the team was heading in the right direction.

Scorecard requirements:

- Studying the market forecast
- Benchmarking our product to others available in the market
- Gathering information on voice of the customer through an online survey
- Translating customer needs to useful metrics and ranking them through building our House of Quality
- Prioritizing customer requirements based on survey results
- Performing Kano analysis
- Completing the House of Quality

4.3.2.1 Identifying "Voice of the Customer"

The team created a survey designed to help understand the most important factors for the tool holder. The team developed this survey specific to the customer. After analyzing these requirements, the team constructed affinity diagrams to rank and structure the customer requirements. Customer requirements fell into three categories:

1. Performance
2. Usability
3. Style

4.3.2.2 Structuring and Ranking Customer Needs

The main wants and needs of the customer were determined through a statistical analysis of the surveys which showed:

1. 87% of the service members who participated in the survey felt that a quick-release detachment would be a great addition to the military tool kit holder.
2. Only 11.8% of service members agree with the current design and believe that no changes need to be made.
3. 20% surveyed made recommendations about increasing ease of use.

4.3.2.3 Analysis of Competitors in the Market

The next step was to identify the competition. Two competitors were identified, a breaching tool holder and a manual entry tool backpack; both utilize existing designs for their tool holders. Neither of these competitors have a design that is unique, with a focus on a functional attachment design.

- Military tool kit holder (cost $23.25)
  - Aluminum tube
  - MOLLE attachment
- Manual Entry Tool Backpack (cost $116.99)
  - Adjustable, nylon backpack

4.3.2.4 Quality Function Deployment

Using the voice of the customer, the team built the House of Quality (HOQ) as shown in Figure 4.2. These are the steps the team used to build the HOQ:

Step 1: List the customer requirements.

Step 2: List the technical descriptors (characteristics that will affect more than one of the customer requirements, in development or production).

Step 3: Compare the two (customer requirements to technical descriptors) and determine relationships.

Step 4: Develop the positive and negative interrelated attributes and identify "trade-offs."

4.3.2.5 Kano Analysis

In order to identify key customer requirements that will make the product more marketable, the team used the Kano model. The Kano model is a

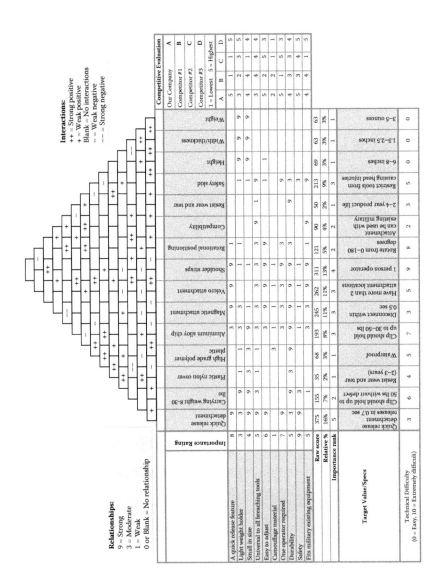

FIGURE 4.2
House of Quality.

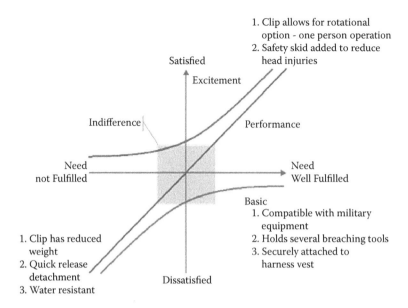

FIGURE 4.3
Kano analysis.

diagram that analyzes customer requirements based on product development as shown in Figure 4.3. These requirements that customers rely on can be put into three categories:

- *Basic Characteristics*: These are features that are not specifically requested but are assumed to be present. If present, the customer is neither satisfied nor dissatisfied. If absent, the customer is very dissatisfied.

- *Performance Characteristics*: These features are specifically requested items from the customer. If present, the customer is satisfied. However, if absent, the customer is dissatisfied.

- *Excitement Characteristics*: These features are unknown to the customer and, therefore, are the most difficult to define and develop. If present, the customer is very satisfied and excited. If absent, the customer is neither satisfied nor dissatisfied.

4.3.3 Define Phase Case Discussion

1. How did your team perform the Voice of the Customer collection? How could VOC collection be improved?

2. Did your team create and distribute a customer survey, and if so, what is the appropriate statistical analysis to perform to identify the importance of the customers' requirements?

3. Did you perform a quality function deployment? How did you identify the technical requirements and the correlations between customer and technical requirements?

4. What is the value of using the Kano model in your VOC analysis?

4.4 Design Phase

4.4.1 Design Phase Activities

1. Identify process elements.
2. Design process.
3. Identify potential risks and inefficiencies.

4.4.2 Design

The process of concept development began with targeting specific issues of the current breaching tool holder system. The main issue discovered was the attachment mechanism. The next step was to begin initial brainstorming and developing redesign concepts for the attachment apparatus. By defining and then aligning the voice of the customer, the design requirements were identified that were necessary to develop an optimal design.

Checklist: Concept generation technique (Theory of Inventive Problem Solving [TRIZ], brainstorming), Design for Manufacture and Assembly, concept generation, affinity diagrams, Pugh concept evaluation and selection

Scorecard requirements:

- Generating seven design concepts that meet customer requirements
- Evaluating superior concepts and superior technology to beat the market
- Analyzing and studying feasibility of superior concepts—looking at the competitors
- Adding value to the design by thinking "outside the box"

The team generated seven design concepts:

- Design 1: Velcro (hook and loop fastener)
- Design 2: Linear clip (rotational)
- Design 3: Magnetic (rotational) attachment
- Design 4: Rotational (multipositional) clip

FIGURE 4.4
Concept Design 1.

- Design 5: Shoulder (backpack) strap
- Design 6: Ball and socket (optimal) attachment
- Design 7: Original design (improve existing PALS system)

4.4.2.1 Design 1: Velcro (Hook and Loop Fastener)

Design 1 is shown in Figure 4.4.

- Characteristics:
  - High-quality, durable hook and loop fastener
  - Two-panel design
    - PALS compatible base (fits current MOLLE equipment)
  - Surface area 8″ × 4″ or 24 in² (recommend 8″ × 8″ or 64 in² hook and loop fastener)
- Limitations:
  - Two-person attachment requirement
  - Velcro wears over time and use
    - Degradation from dirt, water, mildew, and so forth
  - Safety, not rotational
    - Does not allow for emergency removal of the tool

4.4.2.2 Design 2: Linear Clip (Rotational)

Design 2 is shown in Figure 4.5.

- Characteristic(s):
  - Linear clip: recommend a polymer material for both male and female ends

FIGURE 4.5
Concept Design 2.

- Spring tension with mechanical release, located at the bottom of the attachment
- Two-panel design
  - PALS compatible base (fits current MOLLE equipment)
- Rotational (180°) or articulating joint that mounts the clip to the tool holder
  - 90° max from centerline of the user's body (perpendicular stop point)
- Limitations:
  - Two-person attachment requirement
  - Risk of breaking linear clip if not stored properly
  - Limited position of the mechanical release, bottom release

4.4.2.3 Design 3: Magnetic (Rotational) Attachment

Design 3 is shown in Figure 4.6.

- Characteristics:
  - Magnetic attachment
    - Magnet attached to tool holder
    - Steel or magnetically polar base
  - Recessed center on magnet (receptacle cavity)
  - Appendage (fits into the magnet cavity), centered on the base
  - Current 360° rotation, or position (ideal rotation, 180°)
  - One-panel design
    - PALS compatible base (fits current MOLLE equipment)

FIGURE 4.6
Concept Design 3.

- Limitations:
  - Two-person attachment requirement
  - Heavier system due to the metallic requirement
  - Requires a very strong magnet
  - Safety hazards:
    – Working near mines (explosives), with magnetic fuses
    – Interference with other metallic objects

4.4.2.4 Design 4: Rotational (Multipositional) Clip

Design 4 is shown in Figure 4.7.

- Characteristics:
  - Rotational clip, with a lever
  - Tension locked, mechanical selection lever

FIGURE 4.7
Concept Design 4.

- One-panel design
  - PALS compatible base (fits current MOLLE equipment)
- Max 90° range (ideal rotation, 180°)
- Limitations:
  - Two-person attachment and adjustment requirement
  - Selector lever and teeth may slip under weight
  - Selector lever is overexposed, increases the potential for system failure if lever breaks
    - Increases potential for snagging, hazard

4.4.2.5 Design 5: Shoulder (Backpack) Strap

Design 5 is shown in Figure 4.8.

- Characteristics:
  - Two nylon straps
  - Two polymer buckles
  - Adjustable
  - Single user, does not require assistance
- Limitations:
  - Not rotational
  - Straps create a potential hazard by disrupting the user's ability to access current equipment attachments on the front of the user's OTV.
  - Safety hazard:
    - May prevent or hinder the user's ability during an emergency removal of OTV

FIGURE 4.8
Concept Design 5.

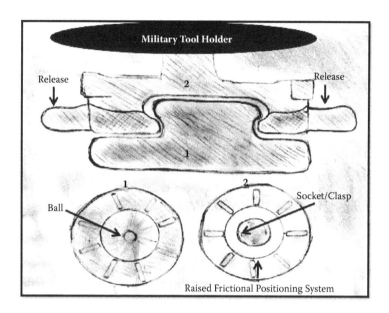

FIGURE 4.9
Concept Design 6.

4.4.2.6 *Design 6: Ball and Socket (Optimal) Attachment*

Design 6 is shown in Figure 4.9.

- Characteristics:
  - Ball and socket, with mechanical release
    - Spring-loaded catch, recessed in the socket
    - Mechanical release located on the side of the mechanism
  - Two circular plates with a ball (1) and socket (2) positioned in the center of each
    - Plate 1 attached to a base that is PALS equipped for attachment to OTV
    - Plate 2 fixed to the military tool kit holder
  - Multiple positioning system
    - Utilizes tension to maintain positioning, along with raised teeth and recess
    - Single user adjustment
  - Aluminum alloy and polymer composition
    - Wear-resistant tension plates
- Limitations:
  - Two-person attachment requirement

4.4.2.7 Design 7: Original Design (Improve Existing PALS System)

- Modifications:
  - Remove PALS system straps from the military tool kit holder and mount a flat nylon panel containing the PALS system
- Use higher-grade nylon webbing straps and snaps

4.4.2.8 Pugh's Concept Selection

The team created seven different designs through our product development. The team then used Pugh's Concept Selection matrix to determine the concept that had the greatest potential for success and to assist in narrowing the scope to a single design as shown in Figure 4.10. The team chose concept Design 7 as the datum design. All other designs are compared to the datum design relative to each customer need. For each comparison, the concept being evaluated is judged to be either better than ("+" score), about the same ("s" score), or worse than the datum ("–" score).

The selected design includes the following feature characteristics:

- Reversal of the plates, improves comfort and prevents snagging from the "ball" appendage
  - Plate 1 (ball) attached to the holder
  - Plate 2 (socket with release) attaches to the MOLLE

- Addition of the safety skid
  - Polymer skid attached to the tool holder's brim
- Prevents injury from the tool accidentally sliding out of the holder

4.4.2.9 Detailed Model of the Product Design

Once the components were identified the team developed a detailed model for the required design specifications as shown in Figure 4.11. (Note that the diagram of the final design is not to scale and is shown in a vertical position. The release tabs are actually perpendicular to the holder.)

4.4.3 Design Phase Case Discussion

1. How did you generate your design concepts?
2. How did you determine how your concepts compared using the Pugh Concept Selection matrix?
3. How did you derive the best combination of your design elements from each concept?

Customer Requirements	Our company	Low enforcement special operation gear	Black hawk	Planned rating	Improvement factor	Sale point	Overall weighting	Current process (Baseline)	Weight/importance	Design 1: Velcro (Hook & loop fastener)	Design 2: Linear clip (Rotational)	Design 3: Magnetic (Rotational) attachment	Design 4: Rotational (Multi-positional) clip	Design 5: Shoulder (Back-pack) strap	Design 6: Ball and socket (Optimal design concept)	Design 7: Original design (No change to existing attachment)
Quick release feature	5	1	0	5	1	1	1	5	1	−	−	S	S	S	S	−
Light weight holder	3	2	3	5	5	1	3	3	8	S	S	−	S	+	+	+
Small in size	4	4	1	4	1	1	1	4	1	S	−	−	S	+	+	−
Universal to all breaching tools	3	4	4	4	3	1	4	3	4	S	S	S	S	S	S	S
Easy to adjust	5	2	5	3	5	1	3	5	3	−	+	S	−	+	S	S
Camouflage material	2	2	0	1	3	1	2	2	2	S	−	−	−	S	S	S
One operator required	5	0	5	3	5	1	3	5	3	+	−	+	+	S	+	−
Durability	4	3	3	4	1	1	1	4	1	−	S	S	+	+	S	+
Safety	5	3	4	5	1	1	1	5	1	−	+	−	+	−	S	+
Fits military equipement	4	4	1	5	3	1	4	5	4	S	+	S	S	+	S	S
Total (+)										1	3	2	3	5	3	3
Total (+)										4	4	4	2	1	0	4
Total (S)										5	3	4	5	4	7	3

FIGURE 4.10
Pugh's Concept Selection matrix.

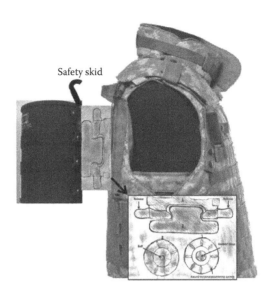

Safety skid

FIGURE 4.11
Final design.

4.5 Optimize Phase

4.5.1 Optimize Phase Activities

1. Implement pilot process.
2. Assess process capabilities.
3. Optimize design.

4.5.2 Optimize

In this phase, the team reviewed the selected superior concept to identify possible flaws and their effects and probable solutions on how to reduce them.

Checklist: Failure mode effect analysis, fault tree analysis
Scorecard requirements:

- Lowering occurrences of high-severity, high-occurrence defects
- Identifying factors influence with one other
- Determining probable solutions to failure

The design can be broken down in different modules (subsystems) according to the functions they perform. A fault tree analysis was performed to identify how all of the modules integrated as shown in Figure 4.12.

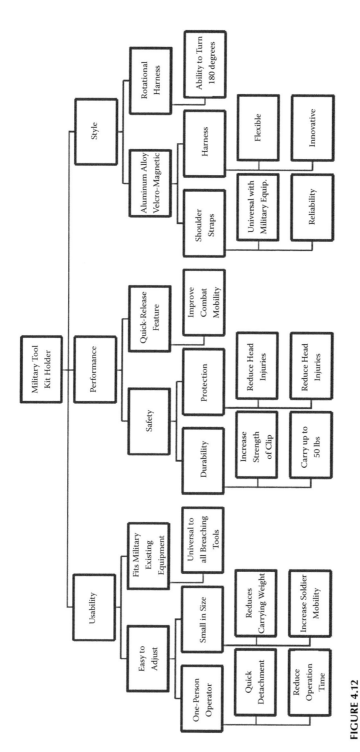

FIGURE 4.12
Fault tree analysis.

4.5.3 Optimize Phase Case Discussion

1. How did you define a failure?
2. How did you determine which factors were significant to your design?
3. How did you determine the design solutions to prevent defects?

4.6 Validate Phase

4.6.1 Validate Phase Activities

1. Validate process.
2. Assess performance, failure modes, and risks.
3. Iterate design and finalize.

4.6.2 Validate Phase

In this phase, the team verified the final design parameters with responses from customers by providing them with a prototype.

Checklist: Measurement system analysis, manufacturing process capability study, reliability assessment, worst-case analysis, analytical tolerance design

Because no quantitative design experiments were conducted in this DFSS project, the team verified that the optimal design requirements met or exceeded the customer requirements (VOC) by showing the prototype to potential customers and users. The customer feedback was very positive, particularly due to the addition of several "excitement" design qualities:

- Performance
  - Safety, durability, multiple positioning (rotational), safety skid
- Usability
- Ease of use, speed, lightweight, MOLLE/PALS compatible

4.6.3 Validate Phase Case Discussion

1. How did you identify potential failure modes of your product?
2. How did you identify potential risks of your product?
3. How would you assess the potential market for your product?

4.7 Summary

Our team utilized the Design for Six Sigma process to successfully develop and design an attachment system that enhances the current military tool kit holder. We are confident that the new attachment will redefine the standards in the breaching tool holder industry and improve the military tool kit holder standing in market sales. The users, specifically those in the military and tactical law enforcement, will have a system that meets their mission-critical equipment needs for durability, ease of use, and safety.

5

Design of a Can Crusher—A Design for Six Sigma Case Study

**Amogh Shenoy, Austin Das, Sagnik Saha, Sumant Joshi,
Elizabeth Cudney, and Sandra Furterer**

CONTENTS

5.1 Project Overview

The product chosen was a portable can crusher. The team plans to design a revolutionary, semiautomatic, robust, paddle-operated can crusher. The can crusher would be designed to crush 16- or 32-ounce cans vertically. The can crusher would have an automatic can feed and drop mechanism for continuous operation with a storage of up to eight cans.

5.1.1 Features of Can Crushers Existing in the Market

- Price was $70 (automatic can crusher costs on the higher end).
- Safety features were provided with medium to high durability.
- Measured size was approximately 2×1 feet.
- Most of them had no rack to store the cans.
- Automatic can crushers weighed around 50 lbs while the manual ones were around 5 to 10 lbs.
- Most were wall mounted with average aesthetics.
- An automatic disposal system was present in most of them.
- They all accepted only a fixed can size.

The aim of the project is to design a portable can crusher that crushes a can to approximately one quarter of its length. The team intends to accommodate two varieties of aluminum cans. It is designed to be eco-friendly and be an ergonomic design. The primary market is household consumers. This can crusher could also be used for light, commercial applications such as restaurants, bars, hotels, motels, and gas stations.

Team requirements and expectations include designing the can crusher based on the voice of the customer (VOC) using Design for Six Sigma (DFSS) tools and techniques to result in a more efficient and organized design process.

The team identified several project boundaries. The first boundary was that the design would only use available technology. The second boundary was the final design should have a price range of $30 to $50.

5.2 Identify Phase

5.2.1 Identify Phase Activities

It is recommended that students work in project teams of three to four students throughout the DFSS case study.

1. *Develop Project Charter*: Use the information provided in the Project Overview section to develop a project charter for the DFSS project.
2. *Team Ground Rules and Roles*: Develop the project team's ground rules and team members' roles.
3. *Develop Project Plan*: Develop your team's project plan for the DFSS project.

5.2.2 Identify

5.2.2.1 Project Charter

The first step was to develop a project charter.

Project Name: Portable garbage crusher system

Project Overview: The goal for the Design for Six Sigma team is to design a portable crusher system that is able to crush cans and plastics (glasses) of comparable size. The crusher would be attachable to a waste basket or recycler with three compartments (i.e., one each for metals, plastics, and glass) and would also have other attachments such as a can opener, a bottle opener, and a drain at the bottom for any leakages or leftover liquids if the cost and design permit. The cost of the system would be approximately $50.

Problem Statement: Today the garbage is usually tossed into the waste container directly, which occupies a lot of space in the waste bin. Also, the current garbage crusher systems available in the market do not have the option to segregate the garbage into metals, plastics, and glass. Using this system the team is planning to build one that can crush the garbage and dispose of it directly into the wastebasket that would be attached. Also, this would be a boost for the recycling effort where companies could provide incentives to households for using this system. In addition, this system would have to be reasonably priced compared to the available ones in the market to be successful in the market.

Customer/Stakeholders: General public, garbage collectors, and the recycling companies. What is important to these customers: ease in use of the system, cost of the system, ergonomic design, and helps in the recycling effort.

Goal of the Project: To build a cheap and portable crusher system.

Scope Statement: The cost of the proposed crusher system has to be almost comparable to the current systems already present in the market, which cost around $30 to $50. Also, there may be a financial benefit for households if the recycling and garbage disposal companies agree to reward the respective households.

Projected Financial Benefit(s): The garbage, as it is already segregated into metals, plastics, and glass, will be much easier to recycle, and if viable, the recycling companies or garbage disposal companies can provide incentives to households that use this system as it makes their work easier.

5.2.2.2 Team Ground Rules and Roles

The team informally developed several ground rules for the project:

- Everyone is responsible for the success of the project.
- Listen to everyone's ideas.
- Treat everyone with respect.
- Contribute fully and actively participate.
- Be on time and prepared for meetings.
- Make decisions by consensus.
- Keep an open mind and appreciate other points of view.
- Communicate openly.
- Share your knowledge, experience, and time.
- Identify a backup resource to complete tasks when not available.

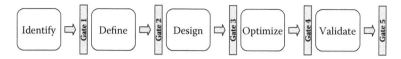

FIGURE 5.1
Design for Six Sigma (DFSS) phase and gate project approach.

5.2.2.3 *Project Plan*

In order to achieve a breakthrough design and successful product commercialization, the team used a structured and disciplined approach of phases and gates in the project plan. The structure consisted of four phases along with four gates. The team also implemented the use of checklists to identify the tools and best practices that were required to fulfill a gate deliverable within a phase. Scorecards were implemented to compare the actual deliverables obtained at each gate versus the required deliverables that were to be accomplished at each gate. The four phases and gates are shown in Figure 5.1.

5.2.2.3.1 *Phase/Gate 1—Identify*

In the Identify phase the team developed the project charter, performed a stakeholder analysis, and created the project plan.

> *Checklist*: Project charter, stakeholder definition, stakeholder analysis, ground rules, project plan

5.2.2.3.2 *Phase/Gate 2—Define*

In this phase, the aim was to research the market and gather customer requirements otherwise known as "voice of the customer." These needs were then translated to useful metrics, prioritized, and evaluated to obtain a clear understanding on customer requirements.

> *Checklist*: Market segment analysis, benchmarking, survey, House of Quality, quality function deployment (QFD), Kano analysis
> - Studying the market forecast
> - Benchmarking our product to others available in market
> - Gathering information on voice of the customer
> - Translating customer needs to useful metrics and ranking them
> - Prioritizing customer requirements
> - Performing Kano analysis
> - House of Quality

5.2.2.3.3 Phase/Gate 3—Design

In this phase, the team developed design concepts with regard to the customer requirements and evaluated each of those concepts to identify the best and most practically deliverable design that satisfies customer needs.

> *Checklist*: Brainstorming, Pugh's Concept Selection, design for X, design for assembly
> - Evaluating superior concepts and superior technology
> - Analyzing and studying feasibility of superior concepts
> - Adding value to the design
> - Generating concepts that meet customer requirements

5.2.2.3.4 Phase/Gate 4—Optimization

In this phase, the team reviewed the selected concept to identify possible flaws and their effects and probable solutions on how to reduce them. The team also compared the influence of one factor on another.

> *Checklist*: Design failure mode and effects analysis (DFMEA), computer-aided engineering (CAE) analysis
> - Identifying noise factors and control factors
> - Lowering occurrences of high-severity, high-occurrence defects
> - Identifying factors that influence one other
> - Identifying probable solutions to failure
> - Performing dimensional optimization

5.2.2.3.5 Phase/Gate 5—Validation

In this phase, the team validated the final design parameters with responses from customers by providing them with prototypes.

> *Checklist*: Virtual prototyping, survey
> - Prototype approved by customer
> - Meets large portion of customer needs
> - Ease to manufacture and cost estimate
> - Reliable performance

A Gantt chart was developed in order to maintain the pace of the project. A detailed timeline for the various tasks within each phase was developed as shown in Figure 5.2. This chart was used to compare the current tasks being carried out to the planned schedule.

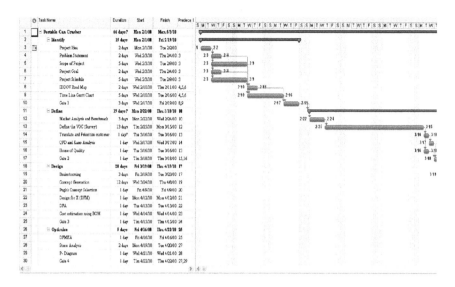

FIGURE 5.2
Project Gantt chart.

5.2.3 Identify Phase Case Discussion

1. DFSS Project Charter: Review the project charter presented.

   a. A problem statement should include a view of what is going on in the business, and when it is occurring. The project statement should provide data to quantify the problem. Does the problem statement provide a clear picture of the business problem? Rewrite the problem statement to improve it.

   b. The goal statement should describe the project team's objective and be quantifiable, if possible. Rewrite the goal statement to improve it.

   c. Did your project charter's scope differ from the example provided? How did you assess what was a reasonable scope for your project?

2. Project Plan

   a. Discuss how your team developed their project plan and how they assigned resources to the tasks. How did the team determine estimated durations for the work activities?

5.3 Define Phase

5.3.1 Define Phase Activities

1. *Collect VOC*: Create a VOC survey to understand current and potential customers' requirements.

2. *Identify critical to satisfaction (CTS) measures and targets*: Based on the VOC, determine the CTS measures and then develop targets using benchmarking data.

3. *Translate VOC into technical requirements*: Using the CTS measures and targets, identify the technical requirements for the product.

5.3.2 Define

In this phase, the aim was to research the recycling and can crusher market and gather customer requirements to capture the VOC through the use of an online survey.

Checklist: Perform market segment analysis by visiting local and online stores and performing a market trend forecast to address the question, "Is there something like this out on the market yet?" In addition, benchmarking, gathering voice of the customer through survey, affinity diagram, quality function deployment, and Kano diagrams were utilized to ensure the team was heading in the right direction.

Scorecard requirements:

- Studying the market forecast
- Benchmarking our product to others available in the market
- Gathering information of voice of the customer through an online survey
- Translating customer needs to useful metrics and ranking them through building our House of Quality
- Prioritizing customer requirements based on survey results
- Performing Kano analysis
- Completing the House of Quality

5.3.2.1 Identifying "Voice of the Customer"

In order to identify the needs of the customer, a survey consisting of essential questions was created. This survey was circulated among various communities with students in the majority. The greatest needs of the customer were identified as shown in Figure 5.3.

	Not Important (%)	No Opinion (%)	Required (%)	Important (%)	Most Important (%)
Portability	48	12	17	16	7
Aesthetics	35	20	24	18	3
Cost	3	3	9	47	38
Durability	3	3	17	39	38
Safety	6	6	16	33	39

FIGURE 5.3
Voice of the customer analysis.

Therefore, from the survey results, it was concluded that the following were the preferred customer needs (in descending order):

- Safe
- Less Cost ($25 to $35)
- Durable
- Portable
- Aesthetically pleasing

Also, customers preferred a can crusher from which some monetary return or incentive could be obtained.

5.3.2.2 Structuring and Ranking Customer Needs

Once the needs were identified the team started with rating those needs so that the focus could be on the most important and critical needs as shown in Figure 5.4.

5.3.2.3 Analysis of Competitors in the Market

The next step was to analyze the current competitors in the market by visiting local department stores and analyzing different brands. Then an extensive Web search was conducted to study the features that are provided on the current products. Five competitors were identified for benchmarking our product (see Figure 5.5).

5.3.2.4 Quality Function Deployment

Based on the voice of customer and competitor's product analysis, a House of Quality was built (see Figure 5.6) as follows:

Step 1: Based on the VOC, the team developed a technical design requirement matrix and then determined the relationship between the customer and engineering requirements.

Number	Needs	Importance	
1	The can crusher	Must be within the range of $20 to $40	5
2	The can crusher	Must have safety provision	5
3	The can crusher	Must have high durability	5
4	The can crusher	Must be easy to use	4
5	The can crusher	Must have low maintenance cost	4
6	The can crusher	Must have a robust mechanism	4
7	The can crusher	Must be able to crush all size aluminum cans	4
8	The can crusher	Should be as automatic as possible	4
9	The can crusher	Allows the provision to store multiple cans	3
10	The can crusher	Allows easy disposal of cans after crushing	3
11	The can crusher	Should have return on investment	3
12	The can crusher	Should be portable	2
13	The can crusher	Should have aesthetic value to it	2

FIGURE 5.4
Ranked customer needs.

Features	Competitor A	Competitor B	Competitor C	Competitor D	Our Company
	High-End Can Crusher	Hand-Operated Can Crusher	Plastic Peddle Operated	Pneumatic Can Crusher	Pedal-Operated Semiautomatic
Price	$250	$20	$25	$200	$30
Safety provision	Provided	Provided	Low	Very low	High
Durability	High	Low	Very Low	High	High
Size (feet)	3×3	1×0.5	1×0.5	3×1	2×1
Mode of operation	Automatic	Manual	Manual	Automatic	Semiautomatic
Rack to store cans	No	No	No	No	Yes
Weight	50 lbs	5 lbs	5 lbs	70 lbs	10 lbs
Portability	Low	High	High	Average	High
Aesthetic	Average	Average	Average	Low	Average
Automatic disposal system	Yes	No	No	Yes	Yes
Maintenance	Very high	Low	Low	Very high	Low

FIGURE 5.5
Competitor analysis.

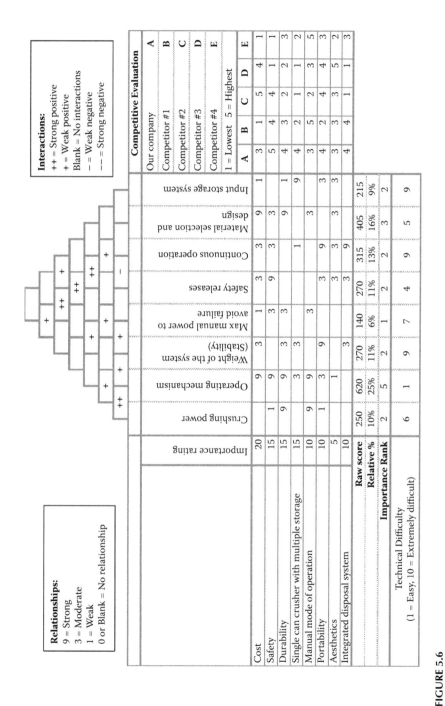

FIGURE 5.6
House of Quality.

Step 2: These requirements were also compared with the products already available in the market.

Step 3: The interrelationship between the technical metrics (engineering characteristics) was studied.

Step 4: Based on the VOC need ranking and technical values, the team determined the difficulty level for each of the design requirements.

5.3.2.5 Product Design Metrics

The next step was to develop the design characteristics metrics to target the ideal specifications that will meet the design requirements as shown in Figure 5.7.

5.3.2.6 Kano Analysis

For the new integrated design of the portable can crusher, the team used the Kano analysis to identify whether the product was meeting the basic customer's one-dimensional needs and expected quality. It was also useful in identifying exciting quality features that could be implemented, which gave the product an advantage over our competitors, and vice versa.

There are certain quality standards that must be implemented and used in conceptual design. These expected qualities that had to be present in the

Number	Design Function	Units	Marginal Value	Ideal Value
1	Maximum crushing force for the can	lbs	200 to 230	250
2	Minimum force applied by a person	lbs	20 to 30	5
3	Maximum height of the can crusher	ft	2 to 4	2
4	Maximum size of the can crusher	sq. ft	2×1	1×1
5	Number of cans for storage	units	6 to 10	12
6	Maximum weight of the can crusher	lbs	10 to 20	10
7	Time required to crush a can	sec	2 to 6	2
8	Deflection of can at maximum load	in	3 to 5	3
9	Reduction from original size of crushed cans	%	25% to 35%	30%
10	Weight of base plate to avoid tipping	lbs	5 to 8	5

FIGURE 5.7
Product design metrics.

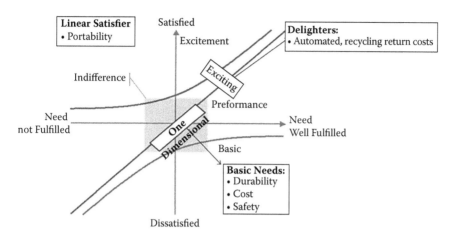

FIGURE 5.8
Kano analysis.

design were high durability of the product, low cost, and high priority for safety. In addition, customers were willing to pay more if portability was an option. The exciting attributes that were also attempted in the design were being a totally semiautomated system and having a recycling return cost from selling crushed cans to recyclers.

Fundamentals: Durability, cost, safety

Linear: Portability

Exciters: Automated, recycling return cost

The Kano analysis is shown in Figure 5.8.

5.3.3 Define Phase Case Discussion

1. How did your team perform the VOC collection? How could VOC collection be improved?
2. Did your team create and distribute a customer survey, and if so, what is the appropriate statistical analysis to perform to identify the importance of the customers' requirements?
3. Did you perform a quality function deployment? How did you identify the technical requirements and the correlations between customer and technical requirements?
4. What is the value of using the Kano model in your VOC analysis?

5.4 Design Phase

5.4.1 Design Phase Activities

1. Identify process elements.
2. Design process.
3. Identify potential risks and inefficiencies.

5.4.2 Design

After passing the Define gate of DFSS, the product enters into the Design phase. Here, the team developed design concepts with regard to the customer requirements and evaluated each of those concepts to identify the best and most practical design that satisfies customer needs.

> *Checklist*: Concept generation technique (Theory of Inventive Problem Solving [TRIZ], brainstorming), design for manufacture and assembly, concept generation, affinity diagrams, Pugh concept evaluation and selection
>
> *Scorecard requirements*:
>
> - Generating seven design concepts that meet customer requirements
> - Evaluating superior concepts and superior technology to beat the market
> - Analyzing and studying feasibility of superior concepts—looking at the competitors
> - Adding value to the design by thinking "outside the box"

5.4.2.1 Concept I

Concept I (shown in Figure 5.9) uses a specially designed cam/follower mechanism to push the piston down to crush the can. The standout feature is the automatic operation (push of a button) that exerts a high crushing force. The disadvantage with this system is that the cam/follower mechanism is expensive to manufacture, which in turn results in the product being very expensive. It should also be noted that power will be needed to drive the cam, which would add cost.

5.4.2.2 Concept II

In Concept II (shown in Figure 5.10) a pneumatic activated piston is used to crush the can. Here a small manual force would be required, and it has a very high crushing force. The disadvantage is the floor space requirement. In addition, a compressor would be needed to actuate the pneumatic pump, which would add cost.

FIGURE 5.9
Concept I.

5.4.2.3 Concept III

In Concept III (shown in Figure 5.11) two power screws are used to convert the rotational movement of the crank/lever into crushing force. The two power screws are interlocked. In addition, a slot is provided for the crushed can to drop down. The disadvantage with this design is that a considerable amount of energy is wasted, and manufacturing the screws would make the design expensive overall. Finally, this crusher needs to be wall mounted, which curbs its portability.

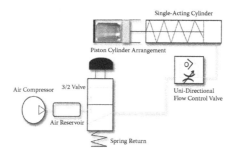

FIGURE 5.10
Concept II.

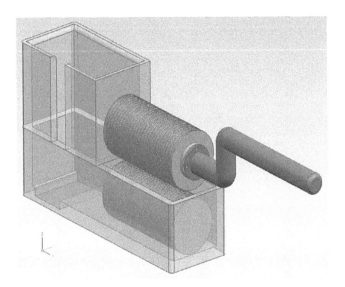

FIGURE 5.11
Concept III.

5.4.2.4 Concept IV

Concept IV (shown in Figure 5.12) is a modification of Concept III. This is a vertical design; therefore, two bevel gears are used in addition to the two power screws. The bevel gears are used to transmit the rotational movement of the lever to the power screws to crush the cans. The disadvantage is that the cost of manufacturing of bevel gears is quite high, which increases the price of the overall product. Also, the can disposal feed must be manual, which can make it unsafe.

5.4.2.5 Concept V

In Concept V (shown in Figure 5.13), only a single horizontal power screw is used to provide the crushing force. The crank/lever is directly connected to the power screw. Also as in all of the other designs, there is a slot provided for the crushed can to drop through. This design is quite energy efficient, but the only disadvantage is that it will take a long time to crush the can, and it is comparatively tedious to operate.

5.4.2.6 Concept VI

The sixth concept (shown in Figure 5.14) was based on a slider concept. Here the crank/lever is connected to the piston or crushing plate with the use of a link. Pulling down on the lever would pull the plate forward, thus crushing the can. This mechanism is quite simple. The disadvantage is that this crusher must be mounted to a wall, which curbs its portability.

FIGURE 5.12
Concept IV.

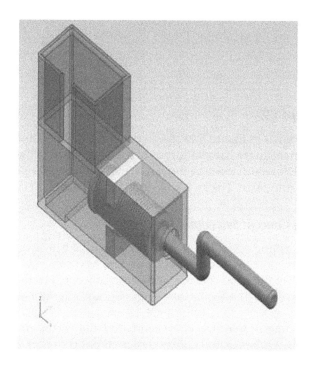

FIGURE 5.13
Concept V.

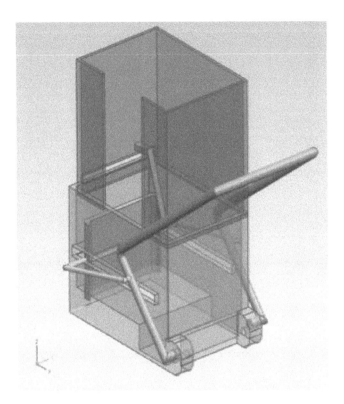

FIGURE 5.14
Concept VI.

5.4.2.7 Concept VII

This design (shown in Figure 5.15) has a pedal for the force to be applied. It also has a mechanism for automatic feed of the cans. Pushing the pedal would move the crank forward crushing the can. When retracting the piston, the next can would fall into place. The rack was designed to hold eight cans.

5.4.2.8 Pugh's Concept Selection

Initial brainstorming for concept generation resulted in seven rough concepts as explained in the previous section. The next step was to rank these concepts to select the ideal concept. A Pugh's Concept Selection matrix was developed to narrow the focus to a single design alternative as shown in Figure 5.16.

This Pugh's concept selection chart compared all the concept designs with Concept Design 1 taken as the datum design. From the chart it was decided that Concept Design 7 was the best choice. Therefore, it was taken as our final design.

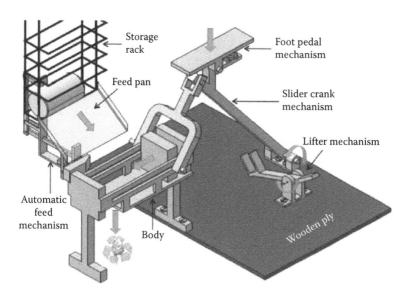

FIGURE 5.15
Concept VII.

5.4.2.9 Concept/Design Development

The design can be decomposed in different mechanisms according to the functions (i.e., slot disposal, retraction mechanism, pedal retaining mechanism, automatic feed mechanism) they perform. The team developed a functional decomposition chart that provides the interrelationship between different modules/subsystems. This chart helped the team to identify the different components necessary to build the final product.

5.4.2.10 Detailed Model of the Product Design

Once the components were identified the team developed a detailed model for the required design specifications as shown in Figure 5.17.

The design actually consists of four mechanisms (shown in Figures 5.18 and 5.19)—that is, the slot mechanism, retraction mechanism, pedal retaining mechanism, and the automatic feed mechanism.

The slot disposal is provided in order to avoid manually removing the crushed can. The crushed can is approximately 30% of the size of the original can. Therefore, the can crusher can be placed right above a recycling bin.

The retraction mechanism is provided to push back the piston to its original location so it is ready to crush the next can.

The pedal retaining mechanism is used to hold the pedal in the correct position (i.e., in the horizontal position). After pressing down on the pedal, the springs return the pedal back to the original position.

Pugh Concept Selection Process Summary Chart

Portable Can Crusher

	DATUM	1	2	3	4	5	6	7
	DATUM	Concept I	Concept II	Concept III	Concept IV	Concept V	Concept VI	Concept VII
Cost	0	−1	−1	−1	−1	−1	1	−1
Safety	0	0	−1	1	1	1	1	1
Durability	0	−1	0	−1	−1	−1	−1	1
Rack for can storage	0	−1	1	1	−1	1	1	1
Mode of operation	0	1	1	−1	−1	0	0	1
Portability	0	−1	0	1	0	1	1	1
Aesthetics	0	−1	1	0	−1	0	1	1
Integrated disposal feature	0	−1	1	0	−1	1	1	1
Number better: S+	+0	+1	+4	+3	+1	+4	+6	+7
Number worse: S−	0	−6	−2	−3	−6	−2	−1	−1
Number same: S0	8	1	2	2	1	2	1	0

FIGURE 5.16
Pugh's Concept Selection matrix.

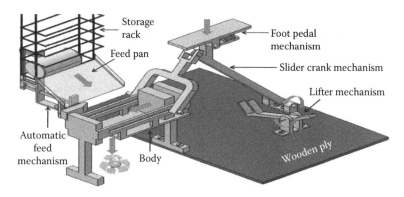

FIGURE 5.17
Final product design.

Designing the automatic feed mechanism was the biggest challenge. The mechanism has been specially designed in such a way that it actuates only when the existing can in the cylinder is crushed to 30% or less of its original size, which ensures that no new can enters the cylinder until the crushed can is disposed of. When the piston crushes the can to 30% the extended dowel pin pushes the bottom link against the springs. This then pushes the top link

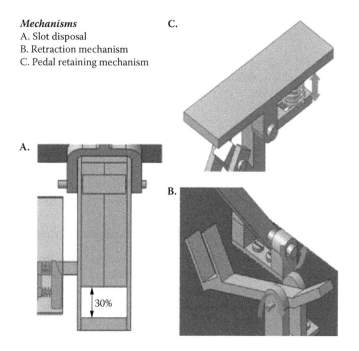

FIGURE 5.18
Mechanisms.

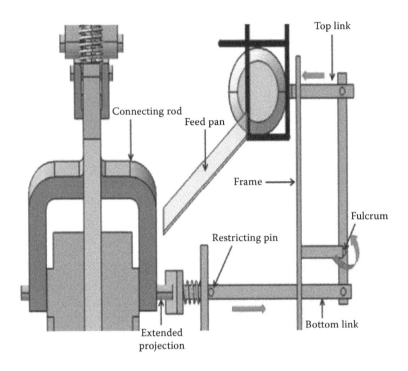

Top link

Connecting rod

Feed pan

Frame

Fulcrum

Restricting pin

Extended
projection

Bottom link

FIGURE 5.19
Automatic feed mechanism.

that is positioned in the rack and displaces and forces the next can down. Once the piston retracts, the mechanism returns to the original position as a result of the spring action.

5.4.2.11 Design for X (DFX)

During the later stages of product development, teams often face difficulties in relating the design requirements to customer needs. This is where DFX steps into action. Design for *X* is a general term where *X* could mean any quality or cost criteria that affect the product. In the project several DFX tools were used.

5.4.2.12 Design for Manufacturing (DFM) and Design for Assembly (DFA)

After finalizing the concept design, it was determined that the components for the product will not be manufactured in house. The components will be outsourced and the product will be assembled in house. A detailed DFA matrix was developed to determine the total assembly time as shown in Figure 5.20. The team prepared the bill of materials (BOM) and the total cost of the product (including the material and labor cost) was determined. Please note that costs have been calculated taking into consideration bulk orders for at least 1000 units.

Number	Part Name	Quantity	*Per Item Cost	Total Cost	Method of Manufacture	Material	Ease of Assembling
1	Dowel pins	9	0.10	0.90	Standard	Standard	Easy
2	Pivot base	2	0.50	1.00	Bought out	6061 Al	Easy
3	Pivot support	4	0.50	2.00	Bought out	6061 Al	Easy
4	Crank	1	4.00	4.00	Bought out	6061 Al	Average
5	Pedal plate	1	1.00	1.00	Standard	6061 Al	Easy
6	Pedal pivot pin	1	1.00	1.00	Standard	Standard	Easy
7	Connecting rod	1	5.00	5.00	Bought out	6061 Al	Difficult
8	Piston	1	2.00	2.00	Bought out	Mild steel	Average
9	Body	1	5.00	5.00	Bought out	6061 Al	Average
10	Spring	3	0.10	0.30	Standard	Standard	Easy
11	Rack	1	2.00	2.00	Bought out	6061 Al	Easy
12	Can feed mechanism	1	2.00	2.00	Bought out	6061 Al	Easy
13	Ply base	1	2.00	2.00	Standard	Wood	Easy
14	Snap rings	22	0.05	1.10	Standard	Standard	Easy
15	Screw	6	0.05	0.30	Standard	Standard	Easy
16	Lifter	1	0.50	0.50	Bought out	6061 Al	Easy
17	Lifter support	1	0.50	0.50	Bought out	6061 Al	Easy

FIGURE 5.20
Design for assembly (DFA) matrix.

Time Required to Assemble Each Part	Seconds
Time to connect the pivot base to pivot	20
Time to couple the crank to pedal	20
Time to connect the pivot pedal pin	20
Time to couple the connecting rod to crank	50
Time to connect the piston to connecting rod	60
Time to assemble the body with piston	30
Time to connect the rack to main body	20
Time to assemble the can feed mechanism	20
Time to mount rack to the body	20
Screw body to the wooden ply base	40
Time to mount lifter assembly on ply base	20
Total time in seconds	320

FIGURE 5.21
Product assembly time matrix.

This matrix helps the team determine how easily these components can be assembled together. It provides the method of fastening for each individual part and gives information if tools are required to assemble the respective part. In addition, the estimated assembly time was determined by the team as shown in Figure 5.21.

The total time for assembly was found to be 5 min 20 sec.

Labor Cost:

Assumed labor cost = $15/hour

This calculated time is an approximate time; therefore, the team added a buffer time.

Therefore, the total time = 6 min/part

Thus, the number of parts manufactured in 1 hour is 10 parts/hour.

Total cost of the materials = $30.60

Total labor cost = $1.50/piece

Total Cost = Material cost + Labor cost = $30.60 + $1.50 = $32.10

5.4.2.13 Design for the Environment

The components used in the product (refer to the BOM) are made from environmentally friendly materials, and all of the parts can be either safely disposed of or recycled for further use.

5.4.3 Design Phase Case Discussion

1. How did you generate your design concepts?
2. How did you determine how your concepts compared using the Pugh Concept Selection matrix?
3. How did you derive the best combination of your design elements from each concept?

5.5 Optimize Phase

5.5.1 Optimize Phase Activities

1. Implement pilot process.
2. Assess process capabilities.
3. Optimize design.

5.5.2 Optimize

In this phase, the team reviewed the selected superior concept to identify the probable problems, major factors involved, and their possible causes. The team also compared the influence of one factor on another and outside noise factors.

Checklist: Design of experiments, failure mode effect analysis, analysis of mean, analysis of variance, design capability studies, critical parameter management

Scorecard requirements:
- Identifying noise factors and control factors from both customers and competitors
- Lowering occurrences of high-severity, high-occurrence defects
- Identifying factors influencing each other
- Determining probable solutions to failure

5.5.2.1 Critical Parameter Management

First, the team created a process diagram with the help of a cause-and-effect diagram to identify the probable problems, major factors involved, and their possible causes. For testing the design, the team next developed a parameter diagram (P-diagram) that identified the control factors, noise factors, and the measurement metrics as shown in Figure 5.22. A P-diagram takes into account all the noise factors and identifies certain control parameters

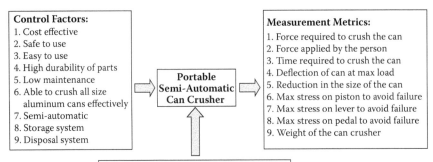

FIGURE 5.22
P-Diagram.

designed to minimize the losses in performance. This results in a more robust design. This P-diagram is then used in preparation for the design failure mode and effects analysis (DFMEA).

5.5.3 Optimize Phase Case Discussion

1. How did you define the Design of Experiment design to use?
2. How did you determine which factors and levels were significant to your design?
3. How did you determine the appropriate number of replications for your experiment?

5.6 Validate Phase

5.6.1 Validate Phase Activities

1. Validate process.
2. Assess performance, failure modes, and risks.
3. Iterate design and finalize.

5.6.2 Validate Phase

In this phase, the team verified the final design parameters with responses from customers.

> *Checklist*: Measurement system analysis, manufacturing process capability study, reliability assessment, worst case analysis, analytical tolerance design.
>
> *Scorecard requirements*:
>
> - Prototype approved by customer
> - Meets large portion of customer needs
> - Ease to manufacture and cost estimate
> - Reliability performance

The DFMEA was developed by first studying the can crusher and brainstorming the possible failures, their consequences, and their severity. Based upon those, the team assigned scores for the probable failures to calculate the risk priority number (RPN). This number was then used as a metric to rank the failures, determine corrective actions, and make recommendations for improvement. The DFMEA is provided in Figure 5.23.

The DFMEA was useful in anticipating probable failures that could arise downstream during production or during customer use. Therefore, it was useful in preventing design failures.

5.6.2.3 Verification Setup

The verification of the product design on Concept VII will be determined in two major parts. The first part will analyze the capability of the product design functional performance. The team will analyze the capability of production, assembly, and manufacturing processes within the business component. The team conducted a design concept evaluation in the university machine shop where it was determined that the design was feasible. This analysis should be straightforward on how to implement current products. For the second part the team conducted a survey of 30 people to anticipate how the public will respond to the product and their satisfaction toward it. The prototype was shown to potential customers, and they were asked if the design met their requirements in the categories of safety, cost, portability, durability, function, and aesthetics. The results are given in Figure 5.24. Based on the survey response, the design was verified to be successful.

5.6.3 Validate Phase Case Discussion

1. How did you identify potential failure modes of your product?
2. How did you identify potential risks of your product?
3. How would you assess the potential market for your product?

Item Number	Item	Function	Potential Failure Mode (Possible Failure)	Potential Failure Effects	Severity	Occurrence	Detection	Current Design Controls	Risk Priority Number (RPN)	Recommended Actions	Severity	Occurrence	Detection	RPN	Percent (%) Reduced
1	Connecting rod	The main function of the connecting rod is to connect the pedal to the piston	Failure to properly analyze the stresses in the rod	Rod breaks due to the stresses	8	7	5	Using steel as a material adding tolerance for the required stresses	280	Use aluminum as the material for connecting rod	7	6	2	84	70%
2	Can feed mechanism	The mechanism feeds the cans to the crushing area one by one	The mechanism gets stuck, becomes loose, or breaks	The cans get stuck in the rack and thus the cans would have to be fed manually	8	7	5	Manually check the joints	280	Evaluate the stability of the entire mechanism before assembling it and also analyze the respective stresses occurring in the links	8	2	2	32	89%

3	Pedal pivot pin	The pedal pivot pin connects the pedal to the piston mechanism	The pivot is not properly connected and designed	The main crushing mechanism would fail due to the application of improper force causing the failure of the entire product	7	Fit the joints tightly with applying the proper tolerances	7	294	Design the pivot pin with adequate support and also tightly fit the pin so as to prevent any motion	7	4	3	84 71%
4	Springs	Springs are used to bring the sliding cranks back to original position	Springs undergo fatigue stress and strain	Due to fatigue in the spring, the mechanism does not come back to the original position, thus affecting the entire mechanism	8	The springs are designed with adequate fatigue tolerances	6	288	Design the springs with adequate stress considerations and fatigue resistances	8	3	1	24 92%

FIGURE 5.23
Design failure mode and effects analysis.

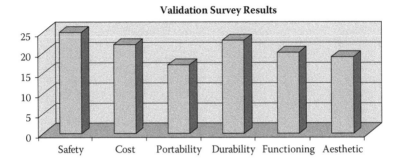

FIGURE 5.24
Validation survey results.

5.7 Summary

Developing a new product platform is a very lengthy and rigorous process. The product design has been shown to account for as much as 80% of the total product cost. Hence, the design process is the most important phase of the overall product development process and, if not managed efficiently, can cost the company additional cost, time, and resources.

The DFSS methodology is key in the design process and when applied effectively results in a robust and reliable design. The tools used and the gating system in DFSS ensure that no design flaws pass to the next phase unnoticed, and the problems are rectified upstream before the product enters production.

As a Six Sigma team, we were able to innovate a commercial portable can crusher with greater customer satisfaction. We were able to gain valuable insight into our potential customers' needs and translate them into a feasible concept that was then redesigned and proved to be commercially viable.

By implementing DFSS throughout the entire conceptualization and redesigning process, we were able to develop a portable can crusher that could be manufactured and assembled with minimal production failures as well as maintaining our overhead and rework costs at a minimum. Overall we were able to meet our project goals within the boundaries and limitations imposed on the project.

6

Design of a Hazardous Chemical Cleanup System—A Design for Six Sigma Case Study

**Benjamin Dowell, James Baker, Prajakta Bhagwat,
Neil Kester, Elizabeth Cudney, and Sandra Furterer**

CONTENTS

6.1 Project Overview

In chemical and hospital laboratories, small liquid spills are a frequent occurrence. Chemical students and medical laboratory personnel use rags or paper towels to mop up these liquids, exposing themselves to chemical or biological risks. Individuals working in these facilities need a product that would cater to their needs of rapid and safer spill cleaning. The current mopping method practiced is time consuming and may be dangerous. With the help of Design for Six Sigma (DFSS) principles, the goal of this team is to design a mop that would absorb approximately 250 mL of chemical or biological liquid spill faster than methods currently in practice and with a stand-off distance of 4 feet ensuring operator safety.

Our goal is to design a cleaning mop to clear chemical and biological liquid spills so as to reduce cleaning time presently required with ordinary mops, thus minimizing exposure to spilled hazardous materials. The mop is provided with a sponge strip to primarily absorb liquid and a synthetic fiber strip to drag up all that remains of the spill. A product with all of these features will approximately cost in the $50 to $60 range. Liquid-specific absorbing sponges would be available in different sizes and capacities.

The product design focuses on

1. Reducing applied force while mopping
2. Providing custom-made sponges withstanding acids and bases, and volatile, flammable, aromatic, body fluids—absorb approximately 200 mL of liquid
3. Discharging mechanism for the used sponge (without touching it)

To assemble our product, the following components are required:

1. Plastic body/staff (4 ft long) with synthetic fiber strip attached at the distal end
2. Synthetic fiber strip (to draw the leftover liquid)

3. Another plastic piece holding the detachable sponge with concentric annuli

4. Spill-specific sponges

The design of the product is based on the voice of the customer (VOC). DFSS will help the design process to be more efficient and organized. The team expects the chemist, school laboratory, medical laboratory communities, and research and development (R&D) laboratories in chemical industries to be the primary markets. The product should yield a 2-year duty life, with the sponges required to be changed after every use and disposed of responsibly. The absorbing sponges are biodegradable and thus are eco-friendly. The team plans to buy spill-specific sponges, fabricate them to fit the product design, and supply them to the customers both wholesale and retail.

6.2 Identify Phase

6.2.1 Identify Phase Activities

It is recommended that students work in project teams of three to four students throughout the DFSS case study.

1. *Develop Project Charter*: Use the information provided in the Project Overview section to develop a project charter for the DFSS project.
2. *Team Ground Rules and Roles*: Develop the project team's ground rules and team members' roles.
3. *Develop Project Plan*: Develop your team's project plan for the DFSS project.

6.2.2 Identify

6.2.2.1 Project Charter

The first step was to develop a project charter.

Project Name: ChemiClean

Project Overview: This product is being considered for production for cleaning of hazardous liquid spills. As part of our enduring efforts toward safer and better laboratory cleaning procedures, this team has considered several options for cleaning of harmful substances. The product (named ChemiClean during the development cycle) is designed to clean small liquid chemical spills faster than methods currently in practice, ensuring operator safety.

Problem Statement: In chemical and hospital laboratories, small liquid spills are a frequent occurrence. Chemical students and medical laboratory personnel use rags or paper towels to mop up these liquids, exposing themselves to chemical or biological risks. Individuals working in these facilities need a product that would cater to their needs of rapid and safer spill cleaning. The current mopping method practiced is time consuming and dangerous.

Customer/Stakeholders: Chemists, school laboratories, medical laboratories, R&D laboratories in chemical industries

Goal of the Project: Reduce time required to clean up liquid chemical spills, and thus minimize exposure to hazardous materials

Scope Statement: The project is limited to the university campus where the team is currently attending classes. The team would conduct a survey along with a field test, among students and medical laboratory personnel at the North Chicago Veterans Affairs Hospital. Product applications and scope of the project are limited to medical and chemical laboratories, due to location restrictions of the student team.

Projected Financial Benefit(s): There are no financial benefits in store for the team while developing this product. However, the consumer will benefit immensely with regard to time spent cleaning up chemical spills coupled with increased cleaning efficiency.

6.2.2.2 Team Ground Rules and Roles

The team informally developed several ground rules for the project:

- Everyone is responsible for the success of the project.
- Listen to everyone's ideas.
- Treat everyone with respect.
- Contribute fully and actively participate.
- Be on time and prepared for meetings.
- Make decisions by consensus.
- Keep an open mind and appreciate other points of view.
- Communicate openly.
- Share your knowledge, experience, and time.
- Identify a backup resource to complete tasks when not available.

6.2.2.3 Project Plan

The project was managed through use of a Gantt chart and a task plan. On establishing the idea/concept (*Invent*), a customer survey was deployed in order to gather customer views. The Gantt chart shows project phases, noting that phases were not completed in parallel, but in series. However, some of the subtasks were completed in parallel. The developed design was optimized and verified using scorecards, which helped in determining which phases could be revisited in order to further improve the product and the process.

A Gantt chart was utilized to maintain the pace of the project and track whether tasks are being carried out per the schedule as shown in Figure 6.1.

6.2.3 Identify Phase Case Discussion

1. DFSS Project Charter: Review the project charter presented.
   a. A problem statement should include a view of what is going on in the business and when it is occurring. The project statement should provide data to quantify the problem. Does the problem statement provide a clear picture of the business problem? Rewrite the problem statement to improve it.
   b. The goal statement should describe the project team's objective and be quantifiable, if possible. Rewrite the goal statement to improve it.
   c. Did your project charter's scope differ from the example provided? How did you assess what was a reasonable scope for your project?
2. Project Plan
   a. Discuss how your team developed their project plan and how they assigned resources to the tasks. How did the team determine estimated durations for the work activities?

6.3 Define Phase

6.3.1 Define Phase Activities

1. *Collect VOC*: Create a VOC survey to understand the current and potential customers' requirements.
2. *Identify critical to satisfaction (CTS) measures and targets*: Based on the VOC, determine the CTS measures and then develop targets using benchmarking data.
3. *Translate VOC into technical requirements*: Using the CTS measures and targets, identify the technical requirements for the product.

Number	Task	Resource	Date Start	Date End	Duration
1	Identify Stage		1/9/2010	2/23/2010	26 days
	Team formation	Team Members	1/19/2010	1/19/2010	1 day
	Define project, develop charter	Team Members	1/19/2010	1/28/2010	8 days
	Charter submittal	Online Resource	2/2/2010	2/2/2010	1 day
	Gathering product information	www.google.com	2/4/2010	2/18/2010	11 days
	Gathering voice of customer	www.surveymonkey.com	2/4/2010	2/18/2010	11 days
2	Define Stage	2/25/2010	3/4/2010	6 days	
	HOQ	Class Slides	2/25/2010	3/4/2010	6 days
	Measurement system analysis	Class Slides	2/25/2010	3/4/2010	6 days
	Process baseline definition	Class Slides	2/25/2010	3/4/2010	6 days
3	Design Stage	3/9/2010	3/18/2010	8 days	
	Kano Analysis	Online Resource	3/9/2010	3/16/2010	6 days
	Analyze Sources of Variation	Online Resource	3/9/2010	3/16/2010	6 days
	Determine Process Drivers	Online Resource	3/9/2010	3/16/2010	6 days

4	Optimize Stage	3/23/2010	4/15/2010	18 days	
	Field Testing	Team Members	3/23/2010	4/6/2010	11 days
	Estimate benefits from new design	Team Members	4/6/2010	4/8/2010	3 days
	Determine and address process	Team Members	4/6/2010	4/8/2010	3 days
	Implement and verify changes	Team Members	4/8/2010	4/15/2010	6 days
5	Verify Stage	4/19/2010	4/29/2010	73 days	
	Document lessons learned	Online Resource	1/19/2019	4/29/2010	73 days
	Final report submission and presentation	Online Resource	4/8/2010	4/29/2010	16 days

FIGURE 6.1
Project Gantt chart.

6.3.2 Define

VOC is a market research technique that produces a detailed set of customer wants and needs, organized into a hierarchical structure, and then prioritized in terms of relative importance and satisfaction with current alternatives. Online surveys were a source for quantitative research, while conversations with potential customers were the primary source for qualitative research. Responses from 15 participants were analyzed and customer needs were sorted into the following categories:

1. Cleaning kit particulars
2. Spills ranked in the order of frequency of confrontation
3. Current competitors

Checklist: Market segment analysis by visiting local stores, performing a market search online, and performing a market trend forecast to address the question, "Is there something like this out on the market yet?" In addition, benchmarking, gathering voice of the customer through survey, affinity diagrams, quality function deployment, and Kano diagrams were utilized to ensure the team was heading in the right direction.

Scorecard requirements:

- Studying the market forecast
- Benchmarking our product to others available in the market
- Gathering information on voice of the customer through an online survey
- Translating customer needs to useful metrics and ranking them through building our House of Quality
- Prioritizing customer requirements based on survey results
- Performing Kano analysis
- Completing the House of Quality

6.3.2.1 Identifying and Ranking the "Voice of the Customer"

After identifying all the wants and needs of the customer, the team ranked and sorted them to realize the most important and critical needs. The VOC was structured and ranked based on customer preferences:

1. Particulars of the cleaning unit (Figure 6.2)
2. Spills ranked in the order of frequency of confrontation (Figure 6.3)
3. Current competitors (Figure 6.4)

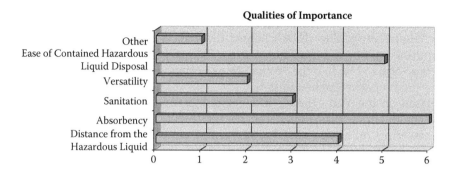

FIGURE 6.2
Cleaning unit customer requirements.

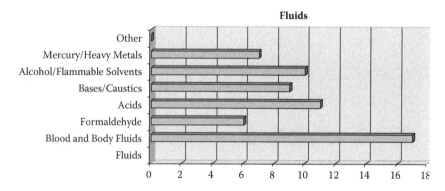

FIGURE 6.3
Rank of chemical usage.

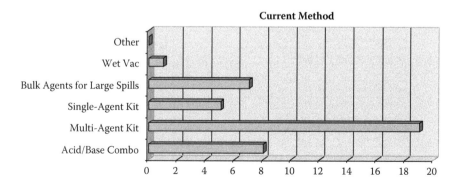

FIGURE 6.4
Current competitors used by customers.

6.3.2.2 Quality Function Deployment

The House of Quality (HOQ) helped the team to transform the VOC into required engineering characteristics for the product as shown in Figure 6.5. This prioritized each of the product characteristics while simultaneously setting product development targets.

Based on the VOC, the team developed a technical design requirement matrix and then determined the relationship between the customer and engineering requirements. The interaction between the technical metrics (engineering characteristics) was studied. Based on the VOC need ranking and technical values, the team determined the difficulty level for each of the design requirements. Finally, the team compared our product characteristics with our competitors in the market.

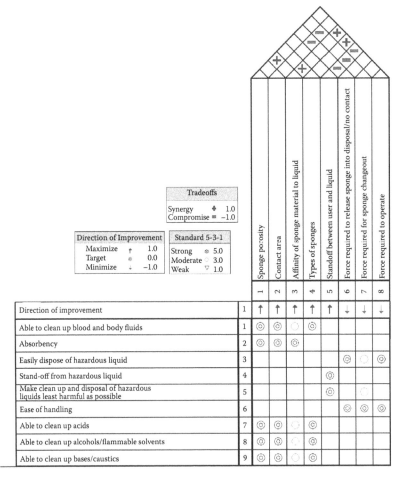

FIGURE 6.5
House of Quality.

6.3.2.3 Kano Analysis

The team used Kano analysis to ensure that customer needs and expectations were met as shown in Figures 6.6 and 6.7. The following set of questions was presented to the initial VOC survey respondents. The exciting quality attributes of the product would be monumental in marketing, competitive pricing, ease of use, and safety.

Questions	Delighted/ Satisfied/Excited Customers	Normally Pleased Customers	Expected/ Basic/Need Well Fulfilled	Majority Rating
If there was a spill cleaner with a stand-off distance of 4.5 ft	1	3	2	Normally pleased
If there was a spill cleaner that cleans acids and bases	1	2	3	Expected
If there was a spill cleaner that cleans acids, bases, and flammable liquids	2	3	1	Normally pleased
If there was a spill cleaner that cleans acids, bases, flammable liquids, and body fluids	4	2	0	Delighted
If there was a spill cleaner that was portable	0	1	5	Expected
If there was a spill cleaner that was durable	1	2	3	Expected
If there was a spill cleaner that was safe	0	0	6	Expected
If there was a spill cleaner that was reusable	1	4	1	Normally pleased
If there was a spill cleaner that was easy to use	0	2	4	Expected
If there was a spill cleaner that was priced below $55	1	2	3	Expected
If there was a spill cleaner priced between $100 and $150	0	0	1	Expected
If the spill cleaner could absorb 250 mL of liquid	1	3	2	Normally pleased
If the spill cleaner could absorb 100 mL of liquid	0	2	4	Expected

FIGURE 6.6
Kano analysis.

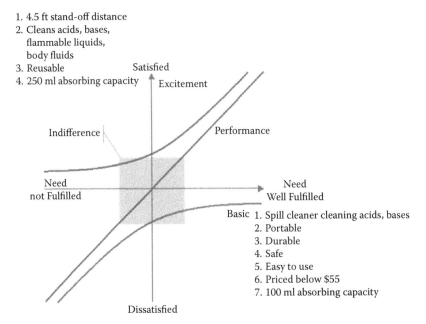

1. 4.5 ft stand-off distance
2. Cleans acids, bases, flammable liquids, body fluids
3. Reusable
4. 250 ml absorbing capacity

Basic
1. Spill cleaner cleaning acids, bases
2. Portable
3. Durable
4. Safe
5. Easy to use
6. Priced below $55
7. 100 ml absorbing capacity

FIGURE 6.7
Kano model.

6.3.3 Define Phase Case Discussion

1. How did your team perform the VOC collection? How could VOC collection be improved?

2. Did your team create and distribute a customer survey, and if so, what is the appropriate statistical analysis to perform to identify the importance of the customers' requirements?

3. Did you perform a quality function deployment? How did you identify the technical requirements and the correlations between customer and technical requirements?

4. What is the value of using the Kano model in your VOC analysis?

6.4 Design Phase

6.4.1 Design Phase Activities

1. Identify process elements.
2. Design process.
3. Identify potential risks and inefficiencies.

6.4.2 Design

At this point the team began the Design phase. During this phase the team took results attained through the HOQ analysis and brainstormed the best ways to effectively and efficiently respond to the technical and customer requirements. During this brainstorming process, each member of the team developed models he or she felt best met those requirements and offered them back to the group for critique and advice. Through this iterative process, the group developed seven designs for the sponge/squeegee configuration and two designs for the ChemiClean main unit. The unit was broken down as shown in Figure 6.8.

> *Checklist*: Concept generation technique (TRIZ, brainstorming), design for manufacture and assembly, concept generation, affinity diagrams, Pugh concept evaluation and selection

Scorecard requirements:

- Generating seven design concepts that meet customer requirements
- Evaluating superior concepts and superior technology to beat the market
- Analyzing and studying feasibility of superior concepts—looking at the competitors
- Adding value to the design by thinking "outside the box"

6.4.2.1 Sponge Configuration

All sponges in these configurations may be composed of multiple types of materials depending on their intended application (see Figure 6.9).

System	Consists of a Working Cleaning Mop Model
Subsystem	1. Liquid absorption/retention of liquid 2. Release mechanism of the sponge 3. Attaching a new sponge piece
Subassembly	1. Release trigger 2. Concentric plastic pipes
Components	1. Sponge 2. Squeegee 3. Lever 4. Plastic body
Manufacturing processes	1. Assembling the pull and squeegee without sponge 2. Specific design and manufacture of sponge

FIGURE 6.8
Subsystem breakdown.

Material	Application
Particulate sorbents	Chemical, acid, and hydrocarbon spills
Sawdust	General application for all types of spills. Sawdust contained/packaged in materials resistant to either acids or bases (two models of this sponge)
Alcohol infused	Biomedical application for sanitization and containment

FIGURE 6.9
Material and application matrix.

Concept I (see Figure 6.10): Provide a straight squeegee 12″ × 1/2″ made of synthetic fiber on the trailing edge of the ChemiClean system (the last portion to come in contact with the contaminated area). This squeegee ensures hazardous liquid consolidates and absorbs into the sponge. Provide a straight sponge 12″ × 2″ that serves as the absorption element in the ClemiClean system.

Concept II (see Figure 6.11): Provide a curved synthetic fiber squeegee 1/2″ thick with a radius of 8″ to the inner edge and located on the trailing edge of the ChemiClean system. Provide a curved sponge 2″ thick with a radius of 5.5″ to the inner edge of the sponge. This design ensures the hazardous liquid is contained within the system and not allowed to escape around the edges.

Concept III (see Figure 6.12): Provide two straight synthetic fiber squeegees 8″ × 1/2″ positioned at a 45° angle on either side of the ChemiClean system. The purpose of these squeegees is to redirect any remaining hazardous liquid into the middle of the ChemiClean system for collection. Provide two 2″ × 8″ sponges positioned at a 45° angle on either side of the ChemiClean system and aligned to the inner side of the squeegees. Provide a third sponge, 3″ × 3″

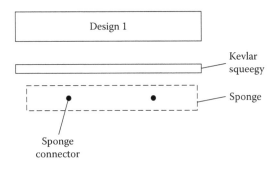

FIGURE 6.10
Concept I.

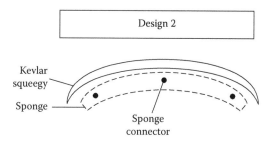

FIGURE 6.11
Concept II.

positioned at a 45° angle to the ChemiClean system and intended to collect any hazardous liquid not absorbed by the first two sponges.

Concept IV (see Figure 6.13): Provide one straight synthetic fiber squeegee 12″ × 1/2″ positioned on the trailing edge of the ChemiClean system. Provide two 12″ × 2″ sponges, one in front of the other on the leading edge of the ChemiClean system. The second sponge is flush with the synthetic fiber squeegee to ensure complete absorption of all remaining hazardous liquid.

Concept V (see Figure 6.14): Provide one straight synthetic fiber squeegee 12″ × 1/2″ positioned on the trailing edge of the ChemiClean system. Provide one sponge shaped as an inverted isosceles triangle with a 12″ base and a height of 8″. This design absorbs an initial quantity of liquid with its leading edge and provides a large surface area along the squeegee's face to ensure complete absorption upon completion of cleaning.

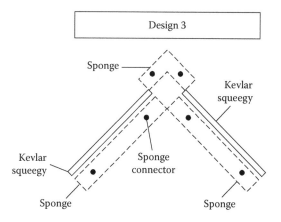

FIGURE 6.12
Concept III.

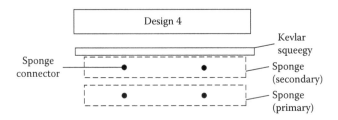

FIGURE 6.13
Concept IV.

Concept VI (see Figure 6.15): Provide a curved synthetic fiber squeegee 1/2″ thick with a radius of 8″ to the inner edge and located on the trailing edge of the ChemiClean system. Provide one circular sponge with a radius of 8″ positioned within the concave portion of the synthetic fiber squeegee.

Concept VII (see Figure 6.16): Provide a synthetic fiber squeegee ring, 1/2″ thick with a diameter of 12″ from inside edge to inside edge. Provide a circular sponge of 12″ diameter with a 2″ diameter section cut out of the center.

6.4.2.2 ChemiClean Main Unit

Concept I (see Figure 6.17): Provide a 4.5-ft-long pole equipped with a handle on the top and a plate on the bottom on which a permanent synthetic fiber squeegee is mounted that has the ability to mount removable sponges. The mechanism used to lock the sponge in place is a spring-loaded hooking mechanism enabled by a wire pulley system. The wire pulley system continues up the pole and is connected

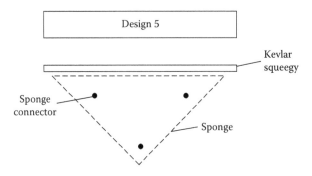

FIGURE 6.14
Concept V.

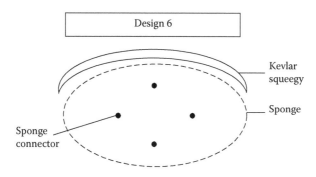

FIGURE 6.15
Concept VI.

to a trigger that disengages the sponge when pulled. This system allows for reclamation and disposal of hazardous liquid while keeping the user 4.5 ft away from the spill.

Concept II (see Figure 6.18): Provide a 4.5-ft-long pole equipped with a foam-gripped handle on one end and a rigid plate on the other. Provide a second pole that surrounds the first and is able to move freely along its length. This second pole has a foam-padded grip on its upper end and a rigid plate on the other. Sponges are attached to the plate on the second pole by snapping the nubs on the top of the sponge into the holes drilled into the second pole's rigid plate. These nubs pass through the first pole's rigid plate without effort because the holes drilled into the first pole's rigid plate are larger in diameter than the sponge's nubs. The sponge is released by holding

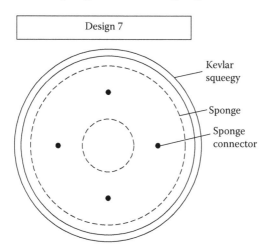

FIGURE 6.16
Concept VII.

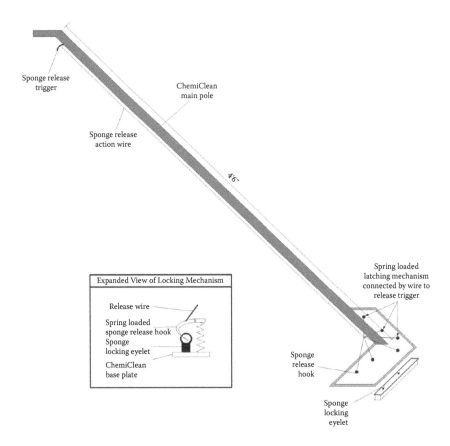

FIGURE 6.17
Main unit Design Concept I.

the second pole in place and pushing down on the first pole. This separates the two plates, forcing the soiled sponge into the designated receptacle.

6.4.2.3 Pugh's Concept Selection

Pugh's Concept Selection matrix helps the team quantitatively rank each concept by assigning a metric that compares that concept to a baseline product. That baseline product may be either a competitor from which you intend on drawing market share or in the case of redesign, your product's earlier model.

As this product is completely new and there are no similar products currently on the market, the team adopted the standard mop as the baseline. The team then developed a list of criteria for comparison based on the voice of the customer used to build the House of Quality. The team then created a spreadsheet (see Figure 6.19) composed of the criteria for

FIGURE 6.18
Main unit Design Concept II.

Pugh's Concept Selection Matrix Criteria for Comparison	Baseline - Standard Mop	Concept 1	Concept 2	Concept 3	Concept 4	Concept 5	Concept 6	Concept 7
Durable (mop head only)	3	-	-	-	-	-	-	-
Portable	5	S	S	S	S	-	-	-
Versatility (effectively clean-up multiple chemicals)	2	+	+	+	+	+	+	+
Distance from liquid when mopping	4	S	S	S	S	S	S	S
Distance from liquid when disposing	1	+	+	+	+	+	+	+
Volumn capable of absorbing	3	S	S	+	+	+	+	+
Ease of mop head change out	1	+	+	+	+	+	+	+
Ease of liquid disposal	1	+	+	+	+	+	+	+
Residue left after mopping	1	+	+	+	+	+	+	+
TOTAL BETTER		5	5	6	6	6	6	6
TOTAL SAME		3	3	2	2	1	1	1
TOTAL WORSE		1	1	1	1	2	2	2
TOTAL POINTS		35	35	37	37	35	35	35

Comparison Key	
Better	+ (5pt)
Same	S (3pt)
Worse	- (1pt)

Baseline Key	
Fully Capable	5
Incapable	1

FIGURE 6.19
Pugh's Concept Selection matrix.

comparison as row headers and the seven concepts and baseline as column headers. Using a scale from 1 to 5 (5 being fully capable and 1 being incapable), the team ranked the baseline for every criteria for comparison. Similarly, the team annotated how each of the concepts (I through VII) compared to the baseline using the criteria: better (+), same (S), and worse (–). Additionally, the team assigned each of these comparisons a point value of 5, 3, and 1, respectively. These values were later used to compare each concept and select the best.

6.4.3 Design Phase Case Discussion

1. How did you generate your design concepts?
2. How did you determine how your concepts compared using the Pugh Concept Selection matrix?
3. How did you derive the best combination of your design elements from each concept?

6.5 Optimize Phase

6.5.1 Optimize Phase Activities

1. Implement pilot process.
2. Assess process capabilities.
3. Optimize design.

6.5.2 Optimize

In this phase, the team explored how the output factors were affected by changes in the input factors. To accomplish this, the team took advantage of several DFSS tools. First, the team used design of experiments to obtain a quantitative assessment of the criticality of key parameters. Next, a parameter diagram was developed to help with the design of experiments (DOE).

Checklist: Design of experiments, failure mode effect analysis, analysis of mean, analysis of variance, design capability studies, critical parameter management

Scorecard requirements:
- Identifying noise factors and control factors from both customers and competitors
- Lowering occurrences of high-severity, high-occurrence defects
- Identifying factors influence with one other
- Determining probable solutions to failure

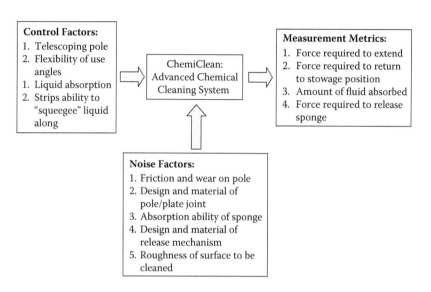

FIGURE 6.20
P-Diagram.

6.5.2.1 Design of Experiments

The team used design of experiments to understand the outputs that were affected by different inputs into the system (see Figure 6.20). This method allowed the team to optimize results in targeted areas based on the results from the voice of the customer and the required functioning parameters. The team determined the most cost-effective way to conduct product development.

The input variables selected were the following: sponge release, extending pole, and spill absorption. Each factor had two levels, high and low, which relate to the input factors being present and absent as shown in Figure 6.21. The output factor of interest is the customer satisfaction rating, 1 to 5. Customers were to assign a satisfaction value for each of the following experimental runs. The data obtained from this experiment were used to compute which factor has the highest influence on customer satisfaction. Overall, it was clear that the absorption capability was the most important, as a certain level was expected. However, the opportunity to deliver highly satisfied results exists also, if greater absorption is combined with the other two features.

6.5.3 Optimize Phase Case Discussion

1. How did you define the DOE design to use?
2. How did you determine which factors and levels were significant to your design?
3. How did you determine the appropriate number of replications for your experiment?

	Customer Value	
	Unhappy	1
	Unsatisfied	2
	Neutral	3
	Satisfied	4
	Highly Satisfied	5

Factors		
Sponge Release	A	
Extending pole	B	
Spill Absorption	C	

Levels	
Present	High
Absent	Low

Run	A	B	C	y1	y2	y3	y	s	s^2
1	low	low	low	1	2	2	1.667	0.577	0.333
2	low	low	high	3	4	2	3.000	1.000	1.000
3	low	high	low	2	2	1	1.667	0.577	0.333
4	low	high	high	3	3	4	3.333	0.577	0.333
5	high	low	low	1	2	2	1.667	0.577	0.333
6	high	low	high	5	4	4	4.333	0.577	0.333
7	high	high	low	2	2	1	1.667	0.577	0.333
8	high	high	high	4	5	5	4.667	0.577	0.333
						Avg	2.750	0.630	0.417

FIGURE 6.21
Design of experiments.

6.6 Validate Phase

6.6.1 Validate Phase Activities

1. Validate process.
2. Assess performance, failure modes, and risks.
3. Iterate design and finalize.

6.6.2 Validate Phase

In this phase, the team verified the final design parameters with responses from customers by providing them with a prototype. The team also identified potential failure modes.

Checklist: Measurement system analysis, manufacturing process capability study, reliability assessment, worst-case analysis, analytical tolerance design

Scorecard requirements:
- Prototype approved by customer
- Meets large portion of customer needs
- Ease of manufacture and cost estimate
- Reliability performance

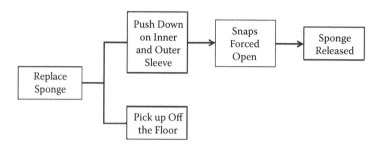

FIGURE 6.22
Functional Analysis System Technique (FAST) tree.

Functional Analysis System Technique (FAST) and Fault tree diagrams were used in the development phase of the project. The FAST tree shows a picture of the important functions of the product, helps ensure a common understanding of the system functions, as well as acts as a tool to envision any hardware solutions or missing components as shown in Figure 6.22.

The Fault tree aids in defining the signal, control, and noise factors that would affect the product, thus helping the team improve the robustness design and identify specific failures that the customer may encounter and appropriate troubleshooting for those failures as shown in Figure 6.23.

6.6.2.1 Design Failure Mode and Effects Analysis

During the failure mode and effects analysis (FMEA) the goal is to quantitatively rank the failure modes and effects for the product. This process allows the team to take preventative action that will improve resistance to failure modes and also to minimize the effects of failure. This FMEA was developed by analyzing the design and figuring out the likely failure scenarios and their effects as shown in Figure 6.24. Two prominent failure modes were identified during this analysis. First, the pole is rigid or not responding to physical input with a risk priority number (RPN) of 28, and second the sponge stuck to plate, with a RPN of 24. These are the highest

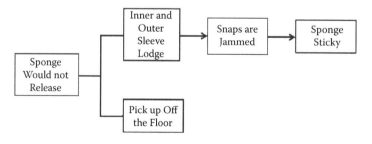

FIGURE 6.23
Fault tree.

Design Failure Modes and Effects Analysis

Item No	Functions	Failure Mode	Effects	Severity	Causes	Occurrences	Controls	Detection	RPN	Recommended Action
1	Absorb spill quickly	Little or no absorption	multiple sponges required or spill not cleaned	9	sponge used beyond shelf life	1	Abide by 2 year shelf life	1	9	None
2	Remote Sponge Release	Sponge stuck to plate	Manual Removal and physical safety	8	Sponge sticks to plate	3	Use non-stick surface on plate	1	24	Look in to non-stick material
		Sponge not released when trigger pulled		8	Snaps jammed	2	Stress / Strain Analysis	1	16	Alter retention / release device
		Sponge not released when trigger pulled		8	Release trigger malfunction	1	Stress / Strain Analysis	1	8	Alter retention / release device
3	Telescoping Pole	Pole fractured	Pole will not telescope out	5	Damage sustained	1	None	1	5	None
		Pole worn		5	Rough or prolonged use	1	Stress / Strain Analysis	3	15	None
		Pole loose		5	Poor assembly	1	Stress / Strain Analysis	2	10	None
4	Squeegee pulls fluid not absorbed	Leaves fluid in trail	Manual Removal and physical safety	6	Spill surface rough	2	None	1	12	None
				6	Damage to Kevlar plate	1	Stress / Strain Analysis	1	6	None
5	Easy to move unit while cleaning	pole rigid or not responding	Discomfort when using	7	Pole-plate junction broken	2	Stress / Strain Analysis	2	28	Use plastic ball and socket joint

FIGURE 6.24
Design failure mode and effects analysis.

RPNs and should be addressed, as these failures disrupt one of the primary requirements as determined by the VOC. The use of a plastic ball and socket would eliminate much of the stress put on this joint and allow for greater flexibility during use. As for the sticky sponge problem, using the product properly can result in this failure. This should be addressed by using a nonstick material or coating on the portion of the plate that comes in contact with the chemical-soaked sponge. This analysis is quite valuable in spotting and correcting potential failures before production has even begun.

6.6.2.2 Design Capability Study

One of the critical parameters as identified in the VOC portion was the amount of liquid that can be absorbed in a single use. The team conducted preliminary research to determine which commercially available sponge should be purchased to mold and fit to the product, and also to determine how much liquid this reformed sponge can absorb. In this capability study the team will determine the amount of liquid absorbed per ChemiClean sponge as shown in Figure 6.25. Recall that the team will be buying off-the-shelf sponges and fitting them to the product.

Product	Dimensions	Area in²	absorb (gal)	absorb per sponge (gal/in²)	absorb per sponge (mL/in²)	Total Fluid Absorbed (mL)	Cost ($)	Cost / gal ($)	Cost per Chemi Clean Pad	Comments
www.thebuyingnetwork.com						Absorbent pads (50 per case)				
GPB50H	15" x 19"	285	14.6	0.00102456	3.879	159.0	$ 23.10	$ 1.58	$ 0.05349	Best Absorbtion
www.thebuyingnetwork.com						Absorbent material (50 ft roll)				
GRB150H	30" x 150'	54000	55.2	0.00102222	3.870	158.7	$ 84.00	$ 1.52	$ 0.05133	Cheapest
www.absorbentsonline.com						Absorbent Pads (Oil Specialty)				
WP100H	15" x 19"	285	33.4	0.00117193	4.436	181.9	$ 41.99	$ 1.26	$ 0.04862	Oil Specific

FIGURE 6.25
Design capability study.

The key metric determined here is the total amount of fluid that a ChemiClean sponge can absorb, denoted by the boxes. This was determined by finding out how much liquid 1 in² of the sponge could absorb, then multiplying it by 33 in², the area of a ChemiClean sponge. The first two differ in absorption by only 3/10 of a milliliter, with the second one being cheaper by 2/10 of a cent. The minimum value required according to the VOC is 100 mL and is easily achieved. However, the 250 mL needed to achieve the excitement factor will require the use of two ChemiClean sponges. Last, the oil specialty sponge seems to perform the best if the spill is a hydrocarbon, as it absorbs the most and is cheaper. Pursuing an oil specialty ChemiClean sponge seems worthwhile as well.

6.6.2.2.1 Cost

As part of this phase, an estimated cost was determined:

Total retail cost: $30
Total manufacturing cost: ~$16
Pole: $10
Sponge: $0.06
Synthetic fiber: $6.00

6.6.2.2.2 Verify Setup

The feasibility of conceptual design, production, assembly, and manufacturing capability of Concept III (sponge) and II (main unit) were verified in this phase. The scorecard for the Concept verification is shown in Figure 6.26.

Also, the components necessary for product manufacture are readily available and are functional to meet product design demands. There would be no time constraints hampering production of this unit. Fulfillment of the above validation phase allows Concept III$_{sponge}$ and II$_{main unit}$ to enter the commercial market (only after filing the copyright laws), and thus be delivered to the customers.

Deliverables	System Performance			Corrective Actions for Problems			Product Design Meets all Requirements		
Conduct final tolerance design on components	R	Y	G	R	Y	G	R	Y	G
Evaluate system components under normal and stress conditions	R	Y	G	R	Y	G	R	Y	G
Run experiments to certify robustness	R	Y	G	R	Y	G	R	Y	G

FIGURE 6.26
Concept verification scorecard.

6.6.3 Validate Phase Case Discussion

1. How did you identify potential failure modes of your product?
2. How did you identify potential risks of your product?
3. How would you assess the potential market for your product?

6.7 Summary

The team implemented the DFSS methodology (Identify-Define-Design-Optimize-Validate, IDDOV) to develop a robust and reliable design of a mop especially for use in chemical and biological liquid spills. The team was able to start with an idea, present it to potential customers, and discovered that they could develop the technology to be robust from an engineering standpoint, and also be robust to unique customer requirements. The phase and gate system of DFSS ensured no design flaws crept into the next phase, and problems could be rectified upstream before the product entered the mass production phase.

The final design consisted of Sponge Design 3 and Pole Design 2. This product answers the VOC through use of HOQ, keeps technicians away from hazardous liquids, and is reusable and portable. The final product is priced at $20 to $30 with a safe stand-off distance of 4 ft, a synthetic fiber strip acting as a squeegee, and replaceable sponge units.

The team was able to develop a mop that could be assembled and manufactured with minimum production failures, simultaneously keeping the overhead and rework costs to a minimum. The team is also prepared with emergency responses to manage mishaps downstream. Thus the IDDOV process helped the team streamline the development, as well as aided in optimizing the design based on voice of the customer; ensuring the project goals were met within the imposed project boundaries and limitations.

7

Design of Radio-Frequency Coverage for Wireless Communications—A Design for Six Sigma Case Study

Roger Ates, Elizabeth Cudney, and Sandra Furterer

CONTENTS

7.1 Project Overview

Wireless handset users have the ability to call into the customer contact center to create a ticket for any network trouble that may be occurring. Customer complaints can range from roaming, poor coverage, to no service in a specific location. The bandwidth frequency of the towers must follow the standards set by the Federal Communications Commission (FCC). The area where the radio frequency (RF) tower is deployed often has no previous data analysis performed to justify the selection of land. As a consequence, the chosen area may not be the most ideal to provide peak performance of the tower, hence creating an unnecessary high cost.

The goal of the project is to determine a new design of directional radio frequency on the current Code Division Multiple Access (CDMA) third-generation (3G) platform and to determine if the added performance of the new design of towers is high enough to justify the extra cost. The redesign goal is to increase the coverage area by 30% and decrease the network coverage trouble tickets by 50%.

The requirements and expectations of the project include the following:

- A design that will be appealing to the voice of the customer (VOC).
- A recommendation of a new design of directional RF on monopole towers for better area coverage or requirements for land a tower is built on.
- An expectation is that by using Design for Six Sigma (DFSS) the customer complaint tickets will drop by 50% to justify redesign.

The project boundary in scope for this project is shown in Figure 7.1 as the device under test (DUT). The scope will only cover customers who are making an in-network mobile-to-mobile call with complete disregard to roaming and intercarrier customer calls.

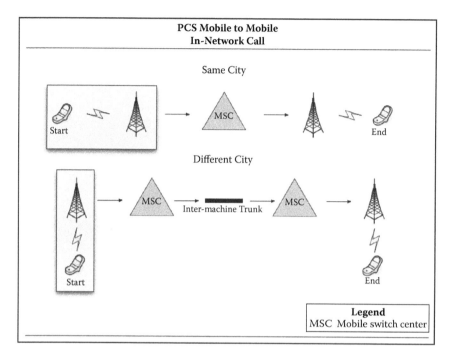

FIGURE 7.1
Project scope diagram.

The tower emits an RF signal in three directions hypothetically covering a strong signal in a 7-mile diameter from the tower. The customer's signal shown on the handset will be dependent on the traffic on the tower and the customer's distance from the tower. It should also be noted that a single sector on the tower can carry up to 20 calls instantaneously, and most towers carry an average of nine sectors.

In order to ensure consistency, the highest traffic complaints area has been chosen to represent the experiment. Figure 7.2 is an example of triangular directional RF emission, which is the standard used on the network infrastructure.

In order to move forward with the evaluated recommendations, the following project boundaries were developed:

- Does not raise cost to customers by more than 5%
- Uses current technology resources and support
- Notes space restrictions
- Follows local zoning laws
- Adheres to FCC regulations

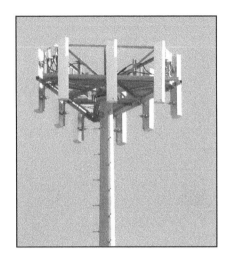

FIGURE 7.2
Example of triangular directional radio-frequency (RF) emission.

7.2 Identify Phase

7.2.1 Identify Phase Activities

It is recommended that students work in project teams of three to four students throughout the DFSS case study.

1. *Develop Project Charter*: Use the information provided in the Project Overview section to develop a project charter for the DFSS project.
2. *Team Ground Rules and Roles*: Develop the project team's ground rules and team members' roles.
3. *Develop Project Plan*: Develop your team's project plan for the DFSS project.

7.2.2 Identify

7.2.2.1 Project Charter

The first step was to develop a project charter.

Project Name: Wireless handset ticket resolution time reduction

Project Overview: The Design for Six Sigma team has set an objective to expand the coverage area of the current RF towers in a major metropolitan area to decrease network trouble tickets for coverage issues. The project aims to reduce by 50% the network trouble tickets generated by customer dissatisfaction and increase the current network coverage by 30%.

Problem Statement: The DFSS team needs to reduce the number of network trouble tickets created by wireless handsets not receiving a satisfactory level of coverage. The trouble tickets need to be reduced by 50% to justify the cost of redesigning the RF towers to increase the coverage by 30% and possibly upgrading to 4G (fourth generation). After analysis of the current state, the team will decide what option to pursue.

The reduction of customer complaints will reduce labor expenses for call centers and RF engineers. The company is looking to be more flexible with the current workforce. Reducing the number of tickets and coverage failures will allow for cross-functional workers to cover more areas of the job. The company will also be able to lease out tower space and coverage to smaller companies, and with a coverage that is more reliable it will be easier to sell.

Customer/Stakeholders: Wireless handset users, smaller wireless companies, stockholders, employees, and suppliers. What is important to these customers (critical to satisfaction, CTS): all of the customers need the company to be successful and profitable.

Goal of the Project: Redesign RF towers to increase coverage area by 30% and reduce network coverage trouble tickets by 50%.

Scope Statement: The scope of this project is RF coverage and reduction in coverage network trouble tickets. The other network tickets are out of the scope of this project.

Projected Financial Benefit(s): This project goal is to reduce operation costs by 10%. These reduced operating costs will be accompanied by increased profits from an increase in customer satisfaction.

7.2.2.2 Team Ground Rules and Roles

The team informally developed several ground rules for the project:

- Everyone is responsible for the success of the project.
- Listen to everyone's ideas.
- Treat everyone with respect.
- Contribute fully and actively participate.
- Be on time and prepared for meetings.
- Make decisions by consensus.
- Keep an open mind and appreciate other points of view.
- Communicate openly.
- Share your knowledge, experience, and time.
- Identify a backup resource to complete tasks when not available.

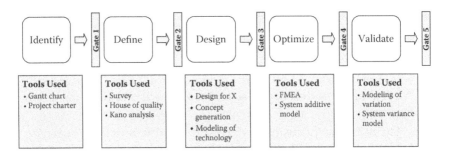

FIGURE 7.3
Identify-Define-Design-Optimize-Validate (IDDOV) phase gate approach.

7.2.2.3 Project Plan

The Identify-Define-Design-Optimize-Validate (IDDOV) methodology was used to apply the DFSS tools. As shown in Figure 7.3, IDDOV is established through a phase gate approach to technology development.

The five phases are as follows:

Phase 1: Identify the problem and develop the project charter.

Phase 2: Define the voice of the customer requirements.

Phase 3: Develop technology concept definition, stabilization, and functional modeling.

Phase 4: Optimize the robustness of the subsystem technologies.

Phase 5: Certify the platform or subsystem technologies.

Each phase should be completed in sequential order. Each uses unique tools to establish the desired answer or result. If completed in sequence, the technique is very effective.

The project began with a discussion as to what the opportunities were for increasing the customer experience with their wireless handset. The conclusion was to create a new RF tower design. This project will be completed using the IDDOV format. An analysis of customer needs, cost, and simplicity should all be considered in the project. This project will be completed with customer needs, design concepts, and modeling of robustness to derive the best possible solution. A Gantt chart was developed to ensure the team kept pace with the tasks required for the successful timely completion of the project.

The final Gantt chart is shown in Figure 7.4 and provides the 15-week progression of the project. The chart was created with the planned and actual timelines based on the preliminary and final estimates. The analysis of the chart shows that Phase 2 was the most time-consuming phase. The team followed the planned time very closely. The team estimated and concluded that the planning and inventing phase would take the longest. The team also found that each phase took a significantly different amount of time. The first

Wireless Handset Gantt Chart in Weeks

Task #	Activity	Planned		Actual		
		Start	Duration	Start	Duration	Done
1	Wireless hand set project	2	15	1	15	100%
Phase 1	**Identify**	2	3	2	3	100%
2	Project goals	2	3	2	3	100%
3	Define expectations	2	3	2	3	100%
Phase 2	**Define**	3	9	3	10	100%
4	Voice of the customer	3	4	3	5	100%
5	Ranking of customer needs	4	5	5	7	100%
6	House of quality	7	7	7	8	100%
7	Kano analysis	8	9	8	9	100%
Phase 3	**Design**	10	12	9	12	100%
8	Design for X methods	10	11	9	10	100%
9	Concept generation	11	12	9	11	100%
10	Modeling of technology	11	12	11	12	100%
Phase 4	**Optimization**	12	13	12	13	100%
11	Modeling of robustness	12	13	12	13	100%
12	System additive model	12	13	12	13	100%
Phase 5	**Verification**	14	15	13	15	100%
13	Modeling of tech platform	14	15	13	15	100%
14	System variance model	14	15	13	15	100%
15	Develop additive model	14	15	14	15	100%

FIGURE 7.4
Project Gantt chart.

two phases were completed along with the schedule, while Phase 3 was completed ahead of schedule and Phase 4 needed an additional week to finish the project on time.

7.2.3 Identify Phase Case Discussion

1. DFSS Project Charter: Review the project charter presented.
    a. A problem statement should include a view of what is going on in the business and when it is occurring. The project statement should provide data to quantify the problem. Does the problem statement provide a clear picture of the business problem? Rewrite the problem statement to improve it.
    b. The goal statement should describe the project team's objective and be quantifiable, if possible. Rewrite the goal statement to improve it.
    c. Did your project charter's scope differ from the example provided? How did you assess what was a reasonable scope for your project?
2. Project Plan
    a. Discuss how your team developed their project plan and how they assigned resources to the tasks. How did the team determine estimated durations for the work activities?

7.3 Define Phase

7.3.1 Define Phase Activities

1. *Collect VOC*: Create a VOC survey to understand the current and potential customers' requirements.
2. *Identify critical to satisfaction (CTS) measures and targets*: Based on the VOC, determine the CTS measures and then develop targets using benchmarking data.
3. *Translate VOC into technical requirements*: Using the CTS measures and targets, identify the technical requirements for the product.

7.3.2 Define

In this project, the customers were already identified as wireless phone users making in-network calls. To obtain the voice of the customer (VOC), an online survey questionnaire was deployed to these customers, and 92 customers responded. An online survey was used because of the low cost and the convenience for the participant to take the survey on his or her own time.

Checklist: Market segment analysis by performing a market trend forecast to address the question, "Is there something like this out on the market yet, and what is considered best in class?" In addition, benchmarking, gathering voice of the customer through a survey, affinity diagrams, quality function deployment, and Kano diagrams were utilized to ensure the team was heading in the right direction.

Scorecard requirements:

- Studying the market forecast
- Benchmarking our product to others available in market
- Gathering information on voice of the customer through an online survey
- Translating customer needs to useful metrics and ranking them through building our House of Quality
- Prioritizing customer requirements based on survey results
- Performing Kano analysis
- Completing the House of Quality

7.3.2.1 Identifying "Voice of the Customer"

To identify the needs of the customer, the team created an online survey designed to help us understand the most important factors to the customer. The responses to the survey are summarized as follows:

1. Rating cell phone service:
   - Reliability 78%
   - Clarity 75%
   - Texting 80%
   - Caller ID 86%
   - Voicemail 78%
2. 56% of respondents were not comfortable with having transmission towers near their residence.
3. 56% of respondents were concerned about the potential dangers of transmission towers.
4. 55% of respondents used their phones for less than 30 minutes in the morning. More people used their phones for less than 2 hours during any part of the day.
5. Respondents had sometimes experienced problems with text messages (sent failed 50%, delayed sent 50%, blocked numbers 12%).

6. 60% did not want to pay more even if the service was improved.

7. Respondents had experienced dropped calls at the following rates: while driving 48%, during inclement weather 35%, and during specific times of the day 29%. However, most experienced dropped calls in amounts they considered acceptable.

8. When asked an open-ended question on how service can be improved, 50 respondents provided an opinion as given below:

   - 18% wanted improved reception/clarity
   - 16% wanted cost to be lowered
   - 10% wanted wider coverage
   - 10% suggested that more transmission towers should be built
   - 8% wanted higher data speed
   - 6% wanted a reduction in dropped calls

Other opinions included having better phones and the freedom to choose their applications, but these were considered to be outside the scope of this project.

7.3.2.2 Structuring and Ranking Customer Needs

Considering all of the customer feedback and the statistical analysis, the team agreed by consensus that the ranked voice of the customer should be as follows (in order):

- Transmission towers, which are far from residences
- Transmission towers, which are safe and perceived so
- Brief but reliable morning call traffic
- Reliable text messaging
- Lower cost
- Reliable calls while driving
- Reliable calls during inclement weather
- Better call clarity

Once the feedback was analyzed, the importance of each need was ranked as shown in Figure 7.5.

7.3.2.3 Analysis of Competitors in the Market

After feedback was analyzed from the customers, the next step was to compare the competitors in the marketplace. Information was gathered by driving around and comparing tower locations, as well as by talking to former employees of the competitors to study the various features. For benchmarking purposes, the top three competitors were identified and the features were compared as shown in Figure 7.6.

	Need		Ranking
1	Radio-frequency (RF) transmission tower	Transmission towers, far from residence	4
2	RF transmission tower	Transmission towers safety	4
3	RF transmission tower	Brief but reliable morning call traffic	3
4	RF transmission tower	Reliable short message service (SMS)	5
5	RF transmission tower	Lower cost of service	2
6	RF transmission tower	Reliable calls while driving	5
7	RF transmission tower	Reliable calls during inclement weather	3
8	RF transmission tower	Faster data connection	2
9	RF transmission tower	Aesthetically appealing transmission towers	1
10	RF transmission tower	Clarity while on a call	5

FIGURE 7.5
Ranked customer needs.

7.3.2.4 Quality Function Deployment

The House of Quality (see Figure 7.7) was built based on the voice of the customer and the competitive analysis. There are four steps that are used in building this House of Quality:

1. Develop a technical design requirement matrix based on the voice of the customer, and determine the relationship between the customer and engineering requirements.

Features	Competitor A	Competitor B	Competitor C
Average tower distance from any residence	500 ft	750 ft	500 ft
Safety requirement according to the National Institute for Occupational Safety and Health (NIOSH)	Below average	Average	Average
Morning call reliability	94%	76%	93%
Short message service (SMS) reliability	86%	92%	88%
Cost of unlimited plan	$129.98/month	$119.98/month	$79.99/month
Data download speed	1.4 Mbps	1 Mbps	1 Mbps
Aesthetic appeal	Average	Average	Average
Clarity of calls placed	Above average	Above average	Average

FIGURE 7.6
Competitor analysis.

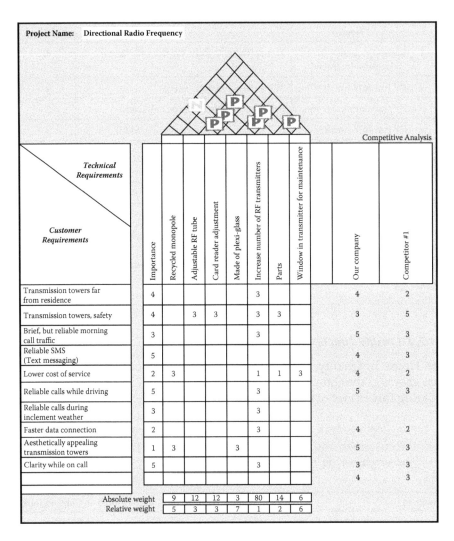

FIGURE 7.7
House of Quality.

2. Compare the requirements to the towers that are already being used in the market.

3. Determine the interrelationship between the engineering characteristics.

4. Establish a difficulty level for each design requirement based on the ranking of the voice of the customer and the engineering characteristic values.

7.3.2.5 Kano Analysis

The Kano analysis model is a customer satisfaction tool used to measure how well a product or service meets customer requirements. The goal is to try and prioritize customer requirements based on how certain attributes satisfy the customer. Based on the previous survey, the team attempted to generate 10 customer satisfaction questions for the Kano analysis. The team talked to 15 participants from three different regions (5 from Wisconsin, 5 from Illinois, and 5 from Missouri) who had taken the previous survey.

Using "Expected" for the basic services customers expect to receive, "Satisfied" for when the customers are satisfied with their current service, "Delighted" for when the customers did not expect to receive the service, and "Not Satisfied" for when the customers deem their service unacceptable. The results (see Figure 7.8) show that a majority of the test market did not approve of living near cell towers and reducing dropped calls would appeal to customers. In addition, the participants were equally delighted and expected to receive better clarity. Finally, none of the participants wanted increased cost for service.

7.3.3 Define Phase Case Discussion

1. How did your team perform the VOC collection? How could VOC collection be improved?

Question	Expected	Satisfied	Delighted	Not Satisfied	Result
A cell tower near your residence	5	2	0	8	Not satisfied
More cell towers built	6	4	5	0	Expected
Service that increased data speed	7	3	5	0	Expected
Service that reduces dropped calls	4	4	7	0	Delighted
Service that provides better reliability	8	3	4	0	Expected
Service that provides better clarity	3	6	6	0	Expected/ delighted
Service that reduces failed sent text messages	7	3	5	0	Expected
Service that eliminates delayed sent messages	4	5	6	0	Delighted
Increased cost of improved service	5	3	0	7	Not satisfied

FIGURE 7.8
Kano analysis.

2. Did your team create and distribute a customer survey, and if so, what is the appropriate statistical analysis to perform to identify the importance of the customers' requirements?

3. Did you perform a quality function deployment? How did you identify the technical requirements and the correlations between customer and technical requirements?

4. What is the value of using the Kano model in your VOC analysis?

7.4 Design Phase

7.4.1 Design Phase Activities

1. Identify process elements.
2. Design process.
3. Identify potential risks and inefficiencies.

7.4.2 Design

After Phase 1, gathering all of the information from the field, the focus was to develop various designs for the transmission towers that may satisfy not only what the customers deem necessary, but also what aligns with the organization's needs and constraints. Three concepts were designed for this project using three-dimensional (3D) modeling programs and other advanced software programs.

Checklist: Concept generation technique (Theory of Inventive Problem Solving [TRIZ], brainstorming), design for manufacture and assembly, concept generation, affinity diagrams, Pugh concept evaluation and selection

Scorecard requirements:

- Generating several design concepts that meet customer requirements
- Evaluating superior concepts and superior technology to beat the market
- Analyzing and studying feasibility of superior concepts—looking at the competitors
- Adding value to the design by thinking "outside the box"

7.4.2.1 Concept 1

The first concept, shown in Figure 7.9, was sculpted around the customer's concern for safety with no cost limit; however, cost was still considered in the use of materials not to exceed the network budget. This design is expected

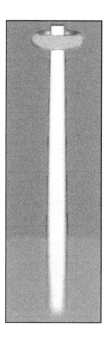

FIGURE 7.9
Concept 1.

to have only a 10% increase in cost of materials to build, but will be cheaper to maintain long term.

The monopole is made from a recycled metal and will have a cut along the sides 180° apart from each other to house a cable system that will be used to adjust the radio-frequency transmitter tube housing up and down the pole not only for engineers to have access to work on the transmitters, but also to allow for easier line-of-sight adjustment of the transmitters. Instead of having access to inside the tower, all that is needed is a card reader and keypad on the outside of the tower to adjust the RF housing as needed.

The RF housing will be made of Plexiglass and will allow up to 20 RF transmitters to be housed. By having more transmitters, the area of coverage can be increased. Plexiglas was chosen because it is a clear, nonreflective material that will not discolor with heat or ultraviolet (UV) light unlike polycarbonate. It is cheap to manufacture and can handle a higher amount of heat; however, to disperse some of the heat four fans will be built into the tube to create air circulation. Each transmitter will have a small window to allow for engineers to be able to work on each one individually and will be opened using the keypad.

7.4.2.2 Concept 2

The second concept, shown in Figure 7.10, was created to not only keep cost in mind, but to also tune into the safety concerns. This design is expected

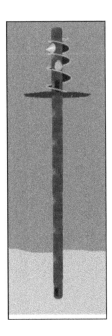

FIGURE 7.10
Concept 2.

to see a 15% drop in material costs, but because a previous structure will be used, maintenance may be needed in the future to keep the structure at a functional level.

For the pole, an old smokestack located near downtown will be used instead of a metal monopole. Inside, a spiral staircase will be used to reach the platform if any maintenance should be necessary on the RF transmitters. A spiral staircase will also be used around the transmitters for easy access.

There will be four cones in a circular direction around the pole, each with three transmitters installed. This will give a circular transmission of radio waves. This pole will have three more transmitters than the current towers allowing for a larger radius of transmission.

7.4.2.3 Concept 3

The final concept, shown in Figure 7.11, was created to keep costs down and to keep a minimalistic design. The design is expected to see a 5% drop in material costs and will use the same monopole design as currently seen on RF towers.

There will be a polyhedral near the top of the pole to house the RF transmitters. Sixteen of the sections in the housing will hold one transmitter. This will allow for more coverage than previously seen in current use of monopoles.

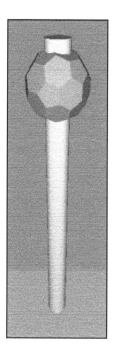

FIGURE 7.11
Concept 3.

This design does not go above and beyond to cover the customer requirement of safety. It uses the same standards of current poles, where a technician is required to climb up the side of the pole or use a lift to reach the transmitters.

7.4.3 Design Phase Case Discussion

1. How did you generate your design concepts?
2. How did you determine how your concepts compared using the Pugh Concept Selection matrix?
3. How did you derive the best combination of your design elements from each concept?

7.5 Optimize Phase

7.5.1 Optimize Phase Activities

1. Implement pilot process.
2. Assess process capabilities.
3. Optimize design.

7.5.2 Optimize

In this phase, the team reviewed the concepts to identify possible flaws and their effects and probable solutions on how to reduce them. The team also compared the influence of one factor on another and outside noise factors.

> *Checklist*: Design of experiments, failure mode effect analysis, analysis of mean, analysis of variance, design capability studies, critical parameter management
>
> *Scorecard requirements*:
>
> - Identifying noise factors and control factors from both customers and competitors
> - Lowering occurrences of high-severity, high-occurrence defects
> - Identifying factors that influence one other
> - Determining probable solutions to failure

7.5.2.1 Design for X (DFX)

Over 70% of a final product's costs are determined during design (Boothroyd and Dewhurst, 2001). Because designs and redesigns involve expenses, most companies increase product cost to recover these expenses. For this project, over 60% of the customers responding to the survey questionnaire had expressed that they were unwilling to pay more even if their services were improved. Additional cost constraints had also been put on the project. To ensure that expenses are kept at a minimum for the project, design for manufacture and assembly (DFMA) methods will be incorporated. Studies have shown that companies that use DFMA methods early in product development, on average, use less resources in product development and also have a shorter time to market.

Using the DFMA guidelines, there will be a deliberate effort on manufacturability by making the product structure less complex. There will be a careful selection of materials with respect to cost over life cycle. In particular,

- These materials will be structurally appropriate and, if possible, locally available.
- Items such as transmission towers will be made shorter in length, where possible, to minimize the quantity of material used.
- At the system level, these towers will also be kept at a minimum.
- The number of parts including fasteners will be minimized to reduce assembly time and cost.

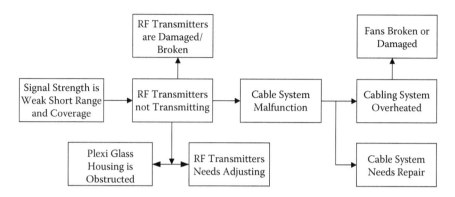

FIGURE 7.12
Functional Analysis System Technique (FAST) tree diagram.

- Parts that are easy to handle and orient during assembly will be produced.
- Improvements will be quantified.

7.5.2.2 Modeling of Technology

The Functional Analysis System Technique (FAST) and Fault tree analysis (FTA) diagrams are used in this phase of the project. The FAST tree diagram, shown in Figure 7.12, is a model outlining the important functions and parts of the project. This helps ensure that everyone understands the product concepts from a physical perspective.

The Fault tree diagram, shown in Figure 7.13, was used to show what noise factors or faults would adversely affect the project. It is used to help identify

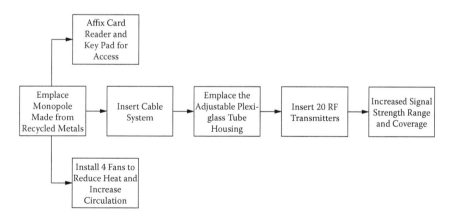

FIGURE 7.13
Fault tree diagram.

what possible failures may occur to help improve robustness. It is also used as a tool to outline troubleshooting.

7.5.2.3 Modeling of Robustness

The failure mode and effects analysis (FMEA) was used to help analyze the potential failure modes within the system and the severity and likelihoods of the failures or defects. The FMEA is the best resource to use that requires minimum effort and cost.

The first step in developing the FMEA for the RF transmission tower was to brainstorm as a group the potential failures and rank the severity of the failure. As shown in Figures 7.14 and 7.15, the scores assigned for the possible failures are used to calculate the risk priority number (RPN), which is then used to determine the rank of severe failures.

7.5.2.4 System Additive Model

In order to provide a robust design, interactions between control factors and noise factors are characterized and quantified as they relate to the critical functional response (CFR). The parameter diagram (P-diagram), shown in Figure 7.16, provides the relationship between the parameters that would affect performance at a system level.

The signal factor in this project is the customer making contact with the tower when calling or sending a message by dialing. This factor is also dependent on the distance from the tower. Call traffic has been included in the noise factors because the team has determined this is a factor they cannot interfere with.

While functional models have not been made for this project, they must be made in the future to fully complete the project. The appropriate signal-to-noise (S/N) ratios will be calculated from the CFR data. In addition, changes may need to be made to the critical parameters to improve the S/N values using design experiments.

The system additive model is generally given by

$$S/N_{opt} = S/N_{avg} + \left(S/N_{A\ opt} - S/N_{avg}\right) + \left(S/N_{B\ opt} - S/N_{avg}\right) + \ldots + \left(S/N_{n\ opt} - S/N_{avg}\right)$$

Either shifting the means or reducing variation or both will optimize the S/N values which will increase the robustness of the system.

7.5.3 Optimize Phase Case Discussion

1. How did you determine which design of experiment to use?
2. How did you determine which factors and levels were significant to your design?
3. How did you determine the appropriate number of replications for your experiment?

Item	Item Function	Potential Failure Mode	Potential Effects of Failure	Severity	Potential Causes of Mechanism Failure
1	Cable for transmitter housing adjustment	1. Cable severed 2. Cable worn 3. Cable fatigue	1. No vertical motion 2. Less stable 3. Play in position	1. 8 2. 7 3. 3	1. Poor-quality material 2. Excessive usage 3. Rough usage
2	Keypad entry	1. Plastic cracks 2. Button stuck 3. System board faulty 4. Indicator light faulty	1. Components exposure to weather 2. Inability to properly use pad 3. Security recognition failure 4. Security recognition failure	1. 2 2. 8 3. 8 4. 1	1. Poor thermal qualities 2. Spring loose 3. Power surge 4. Poor connection
3	Radio-frequency (RF) transmitter	1. Poor connections 2. Antenna loose 3. Power input short	1. Weak signal 2. Intermittent signal 3. Power outage to transmitter	1. 4 2. 6 3. 8	1. Poor-quality material 2. Excessive adjustment 3. Poor-quality materials

FIGURE 7.14
Failure mode and effect analysis (FMEA) for radio-frequency (RF) transmission tower.

Product Failure Mode Effects Analysis—Design FMEA
Product: Radio-Frequency (RF) Transmission Tower

Occur	Current Design Control/Prevention	Current Design Control/Detection	Detection	RPN	Recommended Action
1. 2 2. 3 3. 3	1. Stress analysis 2. Stress/strain analysis 3. None	1. Fatigue test 2. Durability test 3. None	1. 2 2. 2 3. 1	1. 32 2. 42 3. 9	1. None 2. Change material 3. None
1. 4 2. 3 3. 1 4. 2	1. Thermal analysis 2. Stress/strain analysis 3. None 4. None	1. Polymer test 2 Fatigue test 3. None 4. None	1. 3 2. 2 3. 1 4. 1	1. 24 2. 48 3. 8 4. 2	1. Change material 2. Do not use spring in design 3. None 4. None
1. 6 2. 7 3. 4	1. Strain analysis 2. None 3. Strain analysis	1. Durability test 2. None 3. Durability test	1. 2 2. 1 3. 2	1. 48 2. 42 3. 64	1. Change material 2. None 3. Change design

FIGURE 7.15
Failure mode and effect analysis (FMEA) for radio-frequency (RF) transmission tower continued.

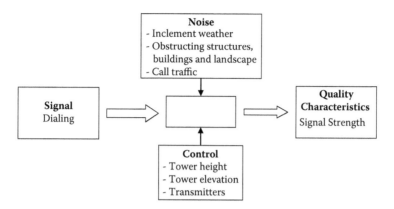

FIGURE 7.16
P-Diagram for the radio-frequency (RF) transmission tower.

7.6 Validate Phase

7.6.1 Validate Phase Activities

1. Validate process.
2. Assess performance, failure modes, and risks.
3. Iterate design and finalize.

7.6.2 Validate Phase

Upon choosing the first design with the recycled monopole and 20 RF transmitters, the team conducted a survey. The survey concluded that over 75% of the participants were more than satisfied with the new tower. The 360° service radius and the increased distance of reception were well-received additions.

Checklist: Measurement system analysis, manufacturing process capability study, reliability assessment, worst-case analysis, analytical tolerance design

Scorecard requirements:
- Prototype approved by customer
- Meets large portion of customer needs
- Ease to manufacture and cost estimate
- Reliability performance

The final step is to create a set of controls to ensure that each tower will be built with the same guidelines. This will include the dimensions and specifications that must be used in construction.

7.6.2.1 System Variance Model

The system variance model is part of the capability verification phase that tests the sensitivities of system tolerance. The system variance model contributes to the parameters of the critical functional response. These results are documented within the system, subsystems, subassemblies, components, and manufacturing process.

In order to conduct a system variance model, the team would need to be able to test the product or service with a test market of consumers. Within this project, that information is not available, and conducting surveys would not be sufficient when trying to determine the system variance of the redesign of cell towers. This type of information would require a field test consisting of a prototype cell tower, cell phones, and consumers. This would enable the team to measure the various noises and tolerances. Due to the time and financial constraints, this was not performed during this case study.

7.6.3 Validate Phase Case Discussion

1. How did you identify potential failure modes of your product?
2. How did you identify potential risks of your product?
3. How would you assess the potential market for your product?

7.7 Summary

In conclusion, the team successfully deployed the use of the IDDOV process within the DFSS methodology to design and develop a new RF transmission tower in the Denver, Colorado, area. By implementing the DFSS process throughout the entire project, the team was able to develop a tower that could be assembled easier, is safer to work on, and gains more coverage area for customers. By using the FMEA, the team was able to become aware of the possible failures and was prepared with recommendations on how to remedy these issues.

The process was very tedious, and most of the time of the project was spent on the design phase of the project. The team was able to take an existing platform and build a new idea to meet the customer requirements. The tools that were used throughout the project and the gating system in DFSS helped to ensure that almost no to minimal flaws were passed on to the next phase unnoticed, and if the flaws do make it to the next phase there is a process in place to rectify them before the product enters production. As a team, we were able to develop a new design for an RF transmission tower that will meet the customer requirements and provide improved customer satisfaction.

References

Boothroyd, G., and Dewhurst, P., *Product Design for Manufacture and Assembly*. New York: CRC Press, 2001.

8

Solar-Heated Jacket—A Design for Six Sigma Case Study

Nick Paul, Josef Garcia, Adam Samiof, Elizabeth Cudney, and Sandra Furterer

CONTENTS

8.1 Project Overview

8.1.1 Project Description

To develop the solar-heated jacket, given the product name "IJacket," a specific methodology, Design for Six Sigma (DFSS), was used to optimize results. DFSS is a business and engineering analysis methodology that uses tools to introduce new technology, products, processes, and services to market. It places the focus up-front in the design or engineering process. The goal is to avoid manufacturing or service process problems using systems engineering techniques from the start. DFSS enables this by focusing the team on understanding the voice of the customer (VOC) and the customer's specific requirements that create detailed engineering specifications. The end result is a robust solution that is optimized to reduce the impact of variation. The product or service, as a result, requires fewer adjustments and thus less money is spent correcting problems. The basic idea is to understand customer requirements before production to reduce costly corrective measures.

8.1.2 Project Description

Currently, there are several jackets that are heated using battery power providing extra warmth as well as a few that can charge portable electronic devices using solar power. The team's idea was to create a jacket that incorporates both of these technologies while still satisfying the expected quality that a winter jacket should fulfill. To ensure the jacket would meet our target market's demand, the team sought customer feedback to guide the process.

8.1.3 Project Goal

The goal of the project is to create a jacket that is capable of providing extra warmth through an interior heating system. It will also be able to provide power to various electronics such as cell phones, iPods, MP3 players, and possibly even larger devices, depending on the VOC. Our aim is to sell at a competitive price point comparable to other winter jackets. Ideally, the IJacket would sell for not much more than a normal winter jacket, making the additional features much more attractive to customers.

8.1.4 Requirements and Expectations

The IJacket will require flexible and durable solar panels that can be attached to the jacket. The solar panels will need to be weather resistant and capable of working in cold temperatures. We expect our customers to find the added technology of the jacket a major delight and key factor in differentiating our product in the market. The jacket should be able to provide constant heating and charging capabilities when in direct sunlight and store up to 3 hours of power when indoors.

8.1.5 Project Boundaries

- The IJacket will use existing technology.
- It will only be able to sustain heating and charging one device at the same time.
- The IJacket will primarily be geared toward outdoor enthusiasts ages 18 to 39.

8.2 Identify Phase Activities

It is recommended that students work in project teams of three to four students throughout the DFSS case study.

1. *Develop Project Charter*: Use the information provided in the Project Overview section to develop a project charter for the DFSS project.
2. *Perform Stakeholder Analysis*: Perform a stakeholder analysis, identifying project stakeholders.
3. *Develop Project Plan*: Develop your team's project plan for the DFSS project.

8.2.1 Identify Phase

8.2.1.1 Project Charter

The first step was to develop a project charter.

Project Name: IJacket

Project Overview: To design a solar-powered jacket with heated interior and a mobile power source.

Problem Statement: There is a recent influx of small electronics carried constantly by individuals in need of a power source. Included in this problem is the fact that many customers use small electronics in the outdoors and likely in cold environments. Heated jackets exist on a 12 V system. However, a greener design using solar energy would be better for the customer. A different jacket exists with a small power source, but its fashion does not appeal to all possible customers. Our expectation is to make the jackets fashionable by incorporating colors and logos of sports teams. We will also include a means of charging small electronics and integrating a sound system for musical enjoyment.

Customer/Stakeholders: We expect the largest group of customers to be outdoor lovers and sports enthusiasts, or possibly even sports teams. The customers would also include those who are constantly traveling.

Goal of the Project: Create a fashionable, green, and affordable solar-powered jacket with all the bells and whistles.

Scope Statement: We will focus on reducing costs to make it more affordable while maintaining the green and fashionable concept.

Projected Financial Benefit(s): Gain the majority of the market share by selling the jackets in the $200 price range.

8.2.1.2 Perform Stakeholder Analysis

The stakeholder analysis was performed to identify the project stakeholders. The stakeholder analysis definition is shown in Figure 8.1.

8.2.1.3 Develop Project Plan

The IJacket project was managed by using a Gantt chart. The Gantt chart is a way to place an expected timeline on the team in order to ensure the project is finished on time. This ensures the product is out to market before the competition in order to gain market share. Although speed is important, speed needs to be balanced with quality assurance measures. To

Stakeholder Name	Stakeholder Role on Project	Impact/Concerns
• Customer	• Key outdoor enthusiasts who would purchase and wear the solar-heated jacket	• Durable jacket • Functional • Relatively low cost • Self-contained heat source
• Project team	• Project team members who will design the jacket	• Apply DFSS tools and methodology • Meet customer requirements
• Project champion	• Project champion, also course instructor who teaches the Design for Six Sigma (DFSS) tools and methodology	• Ensure the teams learn and apply the DFSS tools and methodology • Ensure the customer requirements are met through the product design

FIGURE 8.1
Stakeholder analysis definition.

accomplish this, we constructed gates at each phase that included using templates and checklists from the DFSS methodology. We also required the approval or counsel of a DFSS expert on each task before moving to the next task or phase. The general phases of the project followed the process of IDDOV (Identify-Define-Design-Optimize-Validate).

8.2.1.4 Gantt Chart

The Gantt chart (shown in Figure 8.2) was simplified for ease of reading and brevity.

All the tasks listed in the Gantt chart were completed in series because of the gate requirements. However, many of the subtasks were completed in parallel by different individuals from the working team. Each individual then brought the completed subtask to a group meeting where the group reviewed and approved the work. Rapid and effective communication is paramount to successful project management. Each member of the team must understand his or her role on the team and the expected inputs that he or she must provide. The environment of our working group allowed for face-to-face communication three to four times a week. The group also had other projects together, which enabled cross talk and periodic progress reports throughout the week. Our group has worked and socialized together before the IJacket project and continued the same group dynamic that produced previous success. Good communication enabled each member of the team to participate in any discussion and provide input on the project. Our methods of communication included face-to-face conversation, written correspondence, and telephone conferences.

Date in Day/Month	2-Jan	26-Jan	9-Feb	23-Feb	9-Mar	23-Mar	6-Apr	20-Apr	4-May
Task									
Receive the assignment	▓								
Develop project charter	▓								
Perform Voice of the Customer		▓	▓						
Perform market analysis					▓				
Create quality function deployment (QFD)						▓			
Complete House of Quality (HOQ)						▓			
Perform concept development							▓		
Develop concept selection/optimization							▓		
Validate final report								▓	
Develop final report								▓	
Present results/final report									▓

FIGURE 8.2
Project plan.

8.2.2 Identify Phase Case Discussion

1. DFSS Project Charter: Review the project charter presented.

   a. A problem statement should include a view of what is going on in the business and when it is occurring. The project statement should provide data to quantify the problem. Does the problem statement provide a clear picture of the business problem? Rewrite the problem statement to improve it.

   b. The goal statement should describe the project team's objective and be quantifiable, if possible. Rewrite the goal statement to improve it.

   c. Did your project charter's scope differ from the example provided? How did you assess what was a reasonable scope for your project?

2. Project Plan

   a. Discuss how your team would develop their project plan and how they assigned resources to the tasks. How would the team determine estimated durations for the work activities?

8.3 Define Phase Activities

1. *Collect VOC*: Create a VOC survey to understand the current and potential customers' requirements.

2. *Identify critical to satisfaction (CTS) measures and targets*: Based on the VOC, determine the CTS measures and then develop targets using benchmarking data.

3. *Translate VOC into technical requirements*: Using the CTS measures and targets, identify the technical requirements for the product.

8.3.1 Define Phase

8.3.1.1 *Collect Voice of the Customer (VOC) Information*

To implement the voice of the customer into the design process, we conducted a survey that was mainly targeted at our key demographic of 18- to 39-year-old outdoor enthusiasts. Essentially, we wanted feedback from the population that would most likely have a need or desire for the IJacket. Some of the activities we saw our target market participating in were attending football games, hiking, skiing/snowboarding, and hunting.

Our survey was conducted online, consisted of nine questions, and was taken by 49 individuals. The respondents were given choices for most questions to help narrow the survey but were given an opportunity toward the end of the survey to identify any features they would desire on a jacket that were not included in the survey.

The key information derived from the survey is as follows:

- 92% of our target audience was represented (i.e., outdoor enthusiasts).
- 47% football spectators, 53% hunting, and 56% skiing/snowboarding were the largest populations.
- 67% wait 3 years or longer to buy a new jacket.
- The average temperature for household thermostats was centered between 68 and 70°F.
- 35% showed a strong interest in the general concept for the IJacket with 73% neutral or better.
- Fit, comfort, and warmth were the most important features of the jacket.
- Team logos were undesirable to customers.
- 67% want to pay less than $250 with the additional features.
- The respondents were willing to pay more for additional features.
- 35% of those who responded to our write-in question stated that numerous/well-positioned pockets were a key feature of a jacket they would own.

After analyzing the raw customer data we developed some essential customer requirements:

- Warm, comfortable, and fit well
- Lots of storage
- Integrate with iPods/MP3 players
- Cost less than $250
- Convertible (adapt to changes in temperatures)
- Durable
- Adjustable

Finally, we translated these requirements into functional requirements (FRs) that could satisfy customer requirements. Some of the important functional requirements were

- Solar-powered heating
- Numerous and secure pockets
- Earbud routing and iPod/MP3 storage compartments
- Removable layers

- Breathable fabric
- Moisture-controlling fabric
- Glove integration

8.3.1.2 Ranking Customer Requirements

After identifying the needs, we needed to next rank the customer requirements and identify their importance. Again, we took the VOC from our survey to understand the importance of each requirement, shown in Figure 8.3.

8.3.1.3 Identify Critical to Satisfaction (CTS) Measures and Targets

8.3.1.3.1 Market Analysis

Our next step was to analyze the competitors in our market against our customer requirements as shown in Figure 8.4. We decided to benchmark our product against an existing battery-powered, solar-powered, and regular winter jacket. Thorough research was conducted to find jackets that were targeting our current market and that were sold near our price point. The solar jacket was the only jacket that was slightly above our target price range, but currently it is the only solar-powered jacket on the market that we could find.

8.3.1.4 Translate Voice of the Customer (VOC) into Technical Requirements

8.3.1.4.1 Quality Function Deployment

Our next step was to use quality function deployment (QFD) to further refine the grouped VOC data. QFD uses the data to develop a clear, ranked set of product development requirements. The House of Quality (HOQ) tool was used as our primary tool for QFD. The HOQ transforms the VOC into technical and business performance requirements. It allows us to understand the correlations between our customer requirements and the functional requirements we developed. The HOQ will show us which requirements have strong relationships with each other and if we have missed fulfilling any customer requirements. It will also show us if we have incorporated a functional requirement that the customer does not desire. The House of Quality is shown in Figures 8.5, 8.6, and 8.7.

8.3.1.4.2 Functional Requirements Relationship (Top of House of Quality)

We next needed to understand the relationships between the functional requirements. Sometimes two functional requirements individually will make your product more robust, but the combination of the two can have a negative impact. For example in our HOQ, we found that increasing solar-powered heating and the number of layers would both add to the comfort of the jacket. However, if both were increased, at a certain point it would

Rank	Customer Requirement	Importance
1	Fit/comfort	5
2	Warmth	5
3	Body surface coverage (keeping forearms, face, and neck covered from cold)	5
4	Storage	5
5	iPod/MP3 integration	4
6	Price (positive indicates an increase in price)	4
7	Temperature regulation versus climate (as temperatures change, jacket is still comfortable)	4
8	Temperature regulation versus individual (different people desire different warmth)	4
9	Convertible (changing jacket by removing features)	4
10	Durability	3
11	Adjustability (adjusting jacket to body)	3
12	Compatibility	3
13	Charging capability	2
14	Style/design	1
15	Climate responsiveness	1

FIGURE 8.3
Customer requirement ranking.

make the jacket too hot and uncomfortable. Improving both at the same time can possibly reduce the comfort and thus reduce the customer satisfaction with the jacket. Through the HOQ we identified a number of relationships between the functional requirements that will be considered when developing concepts.

8.3.1.4.3 Functional Requirement Targets

Based on the VOC and the existing products in the market, we established target limits for each of the functional requirements. To facilitate designing concepts we included the difficulty of each target, which was primarily based on the added cost that each requirement would add to the jacket.

8.3.1.4.4 Kano Analysis

To further understand the VOC, we next conducted a Kano analysis as shown in Figure 8.8. The Kano analysis breaks customer satisfaction into three categories: basic needs that must be in the future product, linear

Customer Requirement	Importance	Competitor A	Battery Powered Heated Jacket	Solar Powered Heated Jacket	1 2 3 4 5
Fit/comfort	5	4	4	4	
Warmth	5	4	3	3	
Body surface coverage (Keeping forearms, face, and neck covered from cold)	5	4	3	4	
Storage	5	3	3	2	
iPod/MP3 integration	4	3	1	5	
Price (Positive indicates an increase in price)	4	3	4	2	
Temperature regulation vs. climate (As temps change jacket is still comfortable)	4	3	4	3	
Temperature regulation vs. individual (Different people desire different warmth)	4	1	4	1	
Convertible (Changing jacket by removing features)	4	5	1	4	
Durability	3	4	3	4	
Adjustability (Adjusting jacket to body)	3	4	2	2	
Compatibility	3	5	3	4	
Charging capability	2	1	2	5	
Style/design	1	4	4	4	
Climate responsiveness	1	4	2	2	

Competitor A	– – – – –
Competitor B	– – – –
Competitor C	————

FIGURE 8.4
Market analysis.

satisfiers that produce proportional satisfaction with their increased presence in the future product, and delighters that the customer is satisfied without but make the product exponentially more appealing. Understanding the customer requirements for the first two categories is extremely important in producing a product that customers would consider purchasing. The final category helps us understand the features that will distinguish our product and promote sales. We established these different categories based on the feedback from our survey and from customer needs defined by what they do; essentially analyzing the popular jackets people are currently buying.

- Expected quality (not requested, but assumed present)
  - Water resistant
  - Strong fabric

Customer Requirements and Functions Requirements Relationship Matrix

Demanded Quality (Customer requirement or whats) \ Quality Characteristics (Functional requirements or hows)	Under arm vents	Water resistant material	Strong fabric	Comfortable fabric	Heated hood/hood	Moisture control	Thickness	Breathable fabric	Removable layers	Glove integration	Retractable hood	Ear bud routing/MP3	Solar powered heated	Secure pockets	Numerous pockets	Solar powered charging
Fit/comfort			△	◎		◎	△	◎	◎	◎				◎		
Warmth					⊘	◎	◎	△		◎				⊘		
Body surface coverage (Keeping forearms, face, and neck covered from cold)				⊘					⊘							
Storage														◎	⊘	
iPod/MP3 integration												⊘		⊘	◎	
Price (Positive indicates an increase in price)			◎			◎	△	◎	◎		◎			◎		⊘
Temperature regulation vs. climate (As temps change jacket is still comfortable)	⊘					◎		◎	⊘				◎			
Temperature regulation vs. individual (Different people desire different warmth)	◎					◎			⊘				◎			
Convertible (Changing jacket by removing features)	◎								⊘		⊘					
Durability		⊘	△					△					△			
Adjustability (Adjusting jacket to body)	⊘										◎					
Compatibility										⊘						
Charging Capability													◎		◎	⊘
Style/design							△						⊘	△		△
Climate responsiveness	◎					◎	◎	◎	⊘	◎	⊘			⊘		

Legend:
- ⊘ = Strong relationship
- ◎ = Moderate relationship
- △ = Weak relationship

FIGURE 8.5
Quality function deployment House of Quality.

- Hood
- Retractable hood
- Secure pockets
- One-dimensional quality (specifically requested)
  - Underarm vents
  - Comfortable fabric
  - Moisture control
  - Thickness

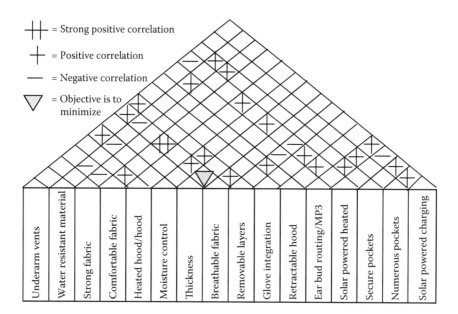

FIGURE 8.6
Top of House of Quality.

- Removable layers
- Numerous pockets
- Exciting quality (unknown to the customer, delighters, wow factor)
  - Heated hood
  - Breathable fabric (for a winter coat)
  - Glove integration
  - Earbud routing
  - Solar-powered heating
  - Solar-powered charging

8.3.2 Define Phase Case Discussion

1. How did your team perform the VOC collection? How could VOC collection be improved?

2. Did your team create and distribute a customer survey, and if so, what is the appropriate statistical analysis to perform to identify the importance of the customers' requirements?

3. Did you perform a quality function deployment? How did you identify the technical requirements and the correlations between customer and technical requirements?

4. What is the value of using the Kano model in your VOC analysis?

Requirement	Target or Limit Value	Difficulty (0 = easy to accomplish; 10 = extremely difficult)	Maximum relationship value in column	Weight/importance	Relative weight
Underarm Vents	Two Vents	0	9	169.8	8.1
Water Resistant Material	Gortex	5	9	45.3	2.1
Strong Fabric	Nylon	5	9	60.4	2.9
Comfortable Fabric	Wool Cotton Blend	5	3	58.6	2.7
Heated Hood/Hood	Cover 80% of Head	5	9	220.8	10.5
Moisture Control	95/5 Polyarmour/Elastane	7	3	84.9	4
Thickness	<2 cm	5	3	47.2	2.2
Breathable Fabric	Polyester Mesh	7	3	94.3	4.5
Removable Layers	One Layer	7	9	271.7	12.9
Glove Integration	Two Thumb Loops	0	9	198.1	9.4
Retractable Hood	Fold Under Less than 3 cm	0	9	107.5	5.1
Earbud Routing/MP3	Pocket iPod/MP3 Compatible	5	9	67.9	3.2
Solar-Powered Heated	70°	7	9	352.8	16.7
Secure Pockets	Double-Stitched Zippers	5	9	107.5	5.1
Numerous Pockets	10 Pockets	5	9	118.9	5.5
Solar-Powered Charging	6 Volts	7	9	103.8	4.9

FIGURE 8.7
Functional requirements targets.

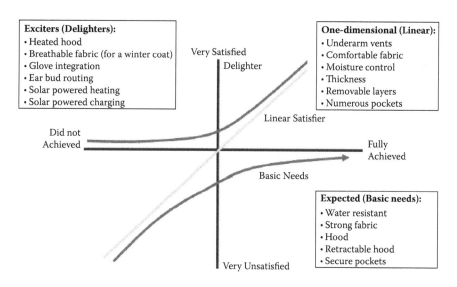

FIGURE 8.8
Kano model for product.

8.4 Design Phase Activities

1. Identify design concepts. Develop the design concepts.
2. Design the product and select the concepts.

8.4.1 Design Phase

8.4.1.1 Identify Design Concepts

8.4.1.1.1 Concept Generation

Concept 1: Our first concept was termed the "Cadillac" option because it would go above and beyond in all facets of quality and capability. The concept is defined by having a strong heating and charging capability. Another distinctive feature is the ability to detach the solar cells from the jacket. In addition, the jacket would have a heated hood, thermostat heating control, iPod/MP3 integration with earbud routing up through the collar, thin and breathable material, lots of pockets, and be composed of multiple layers.

Concept 2: Our second concept was termed the "Extreme Conditions" option, because it would be appropriate for someone who enjoys activities such as mountain climbing where weather extremes are experienced. This concept is defined by having the most powerful solar cells in order to provide large amounts of heat. In addition, the jacket would incorporate glove integration, jacket–pants integration, thick and durable material, rotary heating control, an integrated face shield, and no iPod/MP3 integration.

Concept 3: Our third concept was termed the "Temperate/Moderate Conditions" option. It was intended to be made for climates where there is not a large swing in the weather conditions. This concept would have smaller solar cells, an on/off switch for heating, a non-heating hood, limited pockets, and no electronic charging capability.

Concept 4: Our fourth concept was termed the "Athletic" option. This concept was designed for people who actively participate in sports. Some of the key features that define this concept are thin, breathable material, comfortable fabric, underarm vents, iPod/MP3 integration with earbud routing, smaller solar cells for less heating capability, rotary heating control, glove integration, and multiple layers.

Concept 5: Our fifth concept was termed the "Styled" option because it was meant for a consumer interested in appearing fashionable and trendy. Some of the key features that define this concept are a retractable hood, charging capability, iPod/MP3 integration with earbud routing, no heating capability, and multiple layers.

Concept 6: Our sixth concept was termed the "Convertible" option. This option was meant to exhibit versatility in its use and capabilities. Some features of this option include a retractable hood, underarm vents, multiple layers, numerous pockets, glove integration, moderate solar heating, and limited charging capability.

Concept 7: Our seventh concept was termed the "Storage" option because it was intended to accommodate a customer who likes to carry lots of electronics. This option contains features such as strong charging capability, glove integration, limited heating capability, iPod/MP3 integration with earbud routing, and numerous pockets.

8.4.1.2 Design the Product: Select the Concept

8.4.1.2.1 Pugh's Concept Selection

In order to begin this process, we started by defining the concept selection criteria to conduct our analysis. Due to the manageable amount of VOC requirements, we decided to use all of them when comparing our seven concepts. The VOC requirements are listed in priority, as indicated by the results of our survey. Next, we chose a best-in-class datum to compare our concepts. We chose to use Competitor 1 from our competitive analysis in our House of Quality as the datum because it was the highest rated compared to our VOC. We then conducted our analysis of the seven concepts using the standard + (better), S (same), − (worse) symbols to denote its comparison with the datum. We then compiled the results from the seven concepts, and the results are shown in Figure 8.9.

Our initial analysis of the results indicated the Cadillac option, Convertible option, and Storage option were the strongest options compared to the datum. We then compared the three options to see differences in the ratings. Many of the higher-ranked requirements shared the same rating among all three with the lower-ranked requirements showing the differences. We were able to further our concept using the price criterion. Our projections of price for these three concepts indicated we could meet the customer requirement for less than $250 with the Convertible option and the Storage option, but not the Cadillac option. We decided to form a hybrid option by making trade-offs using the voice of the customer and the price point we intended to meet. Our superior concept contains the following features and is pictured in Figure 8.10:

- Moderate-to-strong heating capability
- Moderate charging capability
- Multiple layers
- Glove integration
- iPod/MP3 integration with earbud routing
- Underarm vents
- Retractable, heated hood
- Numerous pockets (internal and external)
- Comfortable, breathable fabric

8.4.1.3 Design for X (DFX) Methods

Design for Producibility: We intend on using commercially available solar cells, because flexible solar cells capable of providing the needed amounts of voltage and wattage for small, personal electronics already exist. Our base jacket production will be outsourced for cost purposes. We will control the flow of materials through efficient management of the supply chain. We will conduct testing to ascertain the most efficient and safe way to connect the solar cell to the base jacket, which will be our primary role in production.

Design for Assembleability: Because our design will integrate solar cells and fabrics, we will assemble the finished product by hand. The flexibility and soft form of the jacket will require a human to place the coils for the heating component at precise locations. Issues pertaining to the number of fasteners to stabilize the heating elements in the jacket and the ease of these motions will be the most challenging but are capable of being overcome.

Pugh's Concept Selection	Concepts						
Voice of the Customer	Cadillac Version	Extreme Conditions Option	Temperate/Moderate Conditions Option	Athletic Option	Styled Option	Convertible Option	Storage Option
Fit/comfort	+	–	S	+	–	S	S
Warmth	+	+	S	S	–	+	+
Body surface coverage	S	+	S	–	–	S	S
Storage	+	S	–	S	–	+	+
iPod/MP3 integration	+	S	–	+	+	S	+
Price	+	S	+	+	+	+	+
Temperature regulation versus climate	+	+	S	S	–	+	+
Temperature regulation versus individual	+	S	–	+	–	+	+
Convertible	+	–	–	S	+	+	–
Durability	–	+	–	S	–	–	S
Adjustability	+	+	S	+	S	+	S
Compatibility	+	+	–	+	–	S	S
Charging capability	+	S	S	S	+	+	+
Style/design	S	–	–	S	+	S	S
Climate responsiveness	+	–	–	S	–	+	S
Total (+)	12	6	1	6	5	9	7
Total (S)	2	5	6	8	1	5	7
Total (–)	1	4	8	1	9	1	1

FIGURE 8.9
Pugh Concept Selection matrix.

I Jacket Prototype

- Two solar panels on the shoulders
- Right outside pocket contains charger for portable devices
- The left outside pocket contains heating controls
- 10 total pockets (6 outside 4 inside)

FIGURE 8.10
Design concept.

Design for Safety: Design for safety will be paramount in our design with the integration of the solar cells and the jacket material. Inherent in our design will be the treatment of the material that will interface with the coils of the heating element with a flame-resistant compound. This coating will prevent the jacket from catching fire or melting on the skin. Our design also incorporates a safety on/off switch in the lower pocket to quickly disable the transmission of heat. This is in addition to our rotary control for normal heating regulation.

Design for Serviceability: The solar cells we are intending on using come with a 3-year warranty. However, upon entering into a contract with the supplier, the warranty may be adjusted given the use of them on a jacket, because there will be greater wear and tear on the solar cells than the manufacturer may have intended. Potential repair of broken solar cells should be easy to accomplish with the connection point to the jacket easy to access.

8.4.2 Design Phase Case Discussion

1. How did you generate your design concepts?
2. How did you determine how your concepts compared using the Pugh Concept Selection matrix?
3. How did you derive the best combination of your design elements from each concept?

8.5 Optimize Phase Exercises

1. Implement pilot process.
2. Assess process capabilities.
3. Optimize design.

8.5.1 Optimize Phase

To optimize the results of the data we obtained, we decided to form a small parameter diagram as shown in Figure 8.11. We quickly realized that solar energy cells would have the greatest impact on the optimization of the jacket. We needed the most valuable solar energy cell for our situation. The largest increase in efficiency would come from optimizing the solar cell because of the high cost of the solar cells and the variability of capabilities.

The customer requirements showed that we needed 12 V of power. This is the power we needed to warm the jacket and provide a charging function for iPod/MP3 integration simultaneously. Our investigation of solar cells determined the price of 12 V of power at approximately $80. Adding this solar cell would place the price of the jacket above the range desired from the VOC.

We decided to do a comparison of the solar cells we could locate on the market. We compared solar cells that varied greatly in terms of power and increased greatly in cost in terms of power. Larger solar cells cost more per voltage as voltage increased. None of the solar cells met all of the criteria. However, after more brainstorming and group discussion, we decided to compare the single solar cells against different combinations of smaller solar

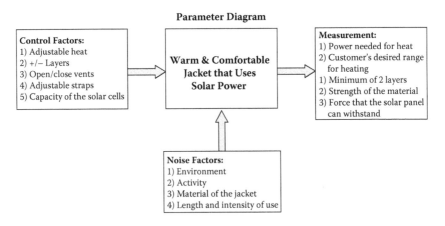

FIGURE 8.11
Parameter diagram.

cells. In this fashion, we successfully identified a combination of smaller solar cells working as separate systems that meet all the customer requirements and kept the total cost of the jacket in the desired range.

8.5.2 Optimize Phase Case Discussion

1. How would you define the design of experiment design to use?
2. How would you determine which factors and levels were significant to your design?
3. How would you determine the appropriate number of replications for your experiment?

8.6 Validate Phase Activities

1. Validate process.
2. Assess performance, failure modes, and risks.
3. Iterate design and finalize.

8.6.1 Validate Phase

The verification of our hybrid super concept must address two key areas before it can be delivered for full production:

1. Ensure that the voice of the customer has been met with the final design.
2. Ensure the voice of the process is feasible with a high degree of consistency and aligns with the expectations of the voice of the customer.

In order to verify that our design meets the voice of the customer, we would propose to conduct a survey to assess the strength with which we met our customers' needs. We would try to get all of the same respondents who filled out our initial survey to reply. Shown in Figure 8.12 is a sample of the areas we would ask to rate on our survey.

We expect the results of this survey to be very positive with averages greater than four. In addition to the survey, we showed our prototype to a few potential customers who are outdoor enthusiasts and would be potential customers for our jacket. Overall, the response to the jacket was very positive with one person remarking, "I would definitely buy this jacket if it were available today!" Therefore, we have successfully met the voice of the customer.

Topics	Rating
	(1 = Not Interested at All; 3 = Neutral; 5 = Very Interested)
Can provide heat to achieve a controlled temperature up to 75°F	
Has a hood with heating capability	
Has retractable hood	
Can charge one small personal electronic in under an hour	
Can hold iPod/MP3 and routes earbuds within the jacket which come out of the collar	
Costs $225	
Has six pockets on the exterior	
Has four pockets on the interior	
Has the ability to connect with winter pants	
Has two underarm vents for temperature regulation	
Has additional internal, optionally removable layer built in	
Is composed of strong and durable material	
Composed of soft, breathable fabric	
How interested are you in buying this?	

FIGURE 8.12
Survey questions.

In the second step of verification, we will ensure the highest level of quality is provided with the finished product. We have a reliable and technologically proficient supplier who can provide the solar cells and a reliable supplier for the base jacket. Our skills and knowledge in connecting the two to provide an actively warming and charging jacket have been established and proven with a high degree of success. Our product has been verified by both the customer and the process.

8.6.2 Validate Phase Case Discussion

1. How would you identify potential failure modes of your product?
2. How would you identify potential risks of your product?
3. How would you assess the potential market for your product?

8.7 Conclusions

The process of developing a new product cannot be undertaken lightly. In order to bring a new product to market, there is a considerable amount of risk involved. Large amounts of capital can be infused into a project with the expectation of positive results. When anything other than those positive results occur, companies are left searching for answers to what went wrong.

DFSS provides a well-conceived methodology to follow in order to reduce and minimize the amount of risk in bringing a new product to market. The cost associated with developing new products can grow tremendously when work has to be redone. The DFSS methodology provides tools and best practices that can be used to limit these potential added costs. The breakdown of the DFSS methodology into phases allows for breaking up the risk into increments, so if something is not progressing with good results, the project can be put on hold or scrapped to minimize the loss. The gates are the points at which the incremental progress is evaluated, and a decision to continue pursuing the project can be made.

In our project, we were able to see this multiphase process in action. We came up with our idea, brought it to our customers to receive their input, and were able to develop a robust and optimized product that met our customers' requirements. We successfully used the IDDOV process to develop this product for our customer. Throughout this whole process, we had a clear vision of what we wanted to accomplish, and we took a methodical approach to reaching it. In the end, we were successful because we kept our customer at the forefront of the process and managed the risks effectively.

9

Traveling Jewelry Box—A Design for Six Sigma Case Study

Robert Jackson, Kayla Najjar, Elizabeth Cudney, and Sandra Furterer

CONTENTS

9.1 Project Overview

9.1.2 Project Description

For this project, the team applied the Design for Six Sigma (DFSS) methodology and tools to redesign the concept of a jewelry case to fit the modern woman's needs while traveling long distances or through her everyday activities. The jewelry box would be designed to provide a convenient travel form that meets her needs for a reasonable price.

9.1.3 Project Goals

Our goal is to design a jewelry case that has the functional capability to carry not only jewelry in a safe and efficient manner but also other items the contemporary woman needs such as makeup, personal items, and even entertainment items, such as an iPod/MP3 or other music player. Consideration to design form will be considered to make the product attractive fashionably, not just functionally. The goal is for women to replace their current jewelry storage products with the new product design. The voice of the customer (VOC) will be implemented to allow the product to attain its maximum value while using DFSS tools and principles in an effort to gain experience in the DFSS process.

9.1.4 Summary of Requirements and Expectations

1. Allocate space for and organize jewelry in a manner that allows for items to be secured (no tangling) and safe from the rigors of travel.
2. Provide a selection of features that distinguish this new jewelry box from other products on the market.
3. Provide previous functions in a fashionable and attractive form.
4. Accomplish the previous requirements in a cost-effective process.
5. Note that the team expects to design a product that meets the customers' needs to an extent that they would be willing to replace their current methods of storing their jewelry while traveling.

9.2 Identify Phase Activities

It is recommended that students work in project teams of three to four students throughout the DFSS case study.

1. *Develop Project Charter*: Use the information provided in the Project Overview section to develop a project charter for the DFSS project.
2. *Perform Stakeholder Analysis*: Perform stakeholder analysis, identifying project stakeholders.
3. *Develop Project Plan*: Develop your team's project plan for the DFSS project.

9.2.1 Identify Phase

9.2.1.1 Project Charter

The first step was to develop a project charter.

Project Name: Traveling Jewelry Box

Project Overview: In the market today, there are very few jewelry boxes that can be considered travel sized and/or reasonably priced. Our team plans to develop such a jewelry box that is appropriately sized for travel (to be determined), that organizes jewelry pieces in a safe environment, that is aesthetically pleasing, and that is considered affordable to the average market (to be determined).

Problem Statement: Our team will set out to create an innovative solution for men and women to be able to transport and have an easy way to display their jewelry without the risk of damaging, tangling, or losing any of the jewelry pieces. This solution will satisfy the needs of the customer because of its capabilities, and for aesthetics and financial concerns as well.

Customers/Stakeholders: Traveling men and women of all ages

Goal of the Project: Create an affordable travel-sized jewelry box.

Scope Statement: Due to a restraint in resources, we will focus on college women. To better understand their requirements, we will survey both the customer and the market to determine the best product design.

Projected Financial Benefit(s): The project will attempt to break into a new market segment for jewelry boxes. If the market is properly analyzed and the customer needs are met and/or exceeded, then as developers we should be profitable.

Project Boundaries: Because the scope of the project is to create an easily portable jewelry case, size will be a constraint. The jewelry case
will have space for only a limited number of items, but its ability to
house different items will need to be flexible. Durable material will
need to be used in its construction over previously used fabrics such
as velvet. The cost of this jewelry box should be under $100 because
it is portable and more likely to be lost than a permanent jewelry box
that would sit in a woman's bedroom.

9.2.1.2 Perform Stakeholder Analysis

The stakeholder analysis was performed to identify the project stakeholders.
The stakeholder analysis definition is shown in Figure 9.1.

The potential customer of the jewelry box would want a jewelry box that
prevents breakage of the items and avoids tangling of necklaces and other
jewelry. The jewelry box should be a reasonable price and be able to be used
at home as well as when traveling. The project team is applying the DFSS
tools and methodology. They desire to meet the customer requirements
through the product design. The project champion, who is also the class
instructor, wants to ensure that the teams learn and apply the DFSS tools
and methodology, and also ensure that the customer requirements are met
through the product design.

9.2.1.3 Develop Project Plan

A Gantt chart was used to keep the project development cycle on track
and ensure that all key steps in the process were accomplished. This chart

Stakeholder Name	Stakeholder Role on Project	Impact/Concerns
• Customer: Women	• Women who carry jewelry while traveling and store it in their homes	• Carry jewelry while traveling without breakage or tangling • Reasonable price • Use at home and sized to travel
• Project team	• Project team members who will design the jewelry box	• Apply DFSS tools and methodology • Meet customer requirements
• Project champion	• Project champion, also course instructor who teaches the Design for Six Sigma (DFSS) tools and methodology	• Ensure the teams learn and apply the DFSS tools and methodology • Ensure the customer requirements are met through the product design

FIGURE 9.1
Stakeholder analysis definition.

includes dates and tasks that are essential for appropriate usage of time management during the scheduled period of work. The team used the Identify-Define-Design-Optimize-Validate (IDDOV) methodology.

9.2.1.3.1 Identify-Define-Design-Optimize-Validate (IDDOV) Roadmap (Plan of Action)

To have a successful product that satisfies a wide array of the target market, project management is key. Using the IDDOV roadmap, the team developed a timeline that included all of the phases and intermediate gates, allowing for a check concluding that all requirements have been met up to that gate.

9.2.1.3.1.1 Phase/Gate 1—Identify This phase is where the team developed the project charter, identified the stakeholders, and developed the project work plan.

9.2.1.3.1.2 Phase/Gate 2—Define This phase was used to learn about what was currently on the market, and what a consumer looks for in a product of this type. Using a carefully constructed survey the voice of the customer was obtained.

> *Checklist*: Gathering voice of the customer, market availability analysis, competitive benchmarking, Kano analysis, quality function deployment (QFD)

> *Scorecard requirements*: Determine customer wants and needs using QFD and Kano analysis, translate survey results to numerical metrics, prioritize customer requirements

9.2.1.3.1.3 Phase/Gate 3—Design This phase was used in the design of several different concepts that met both customer requirements and manufacturing requirements. Each concept has different trade-offs that have to meet customer requirements or manufacturing trade-offs.

> *Checklist*: Generate concepts using what is currently on the market combined with data gathered from the voice of the customer. Use Pugh Concept Selection matrix to select the best elements of the concepts.

> *Scorecard requirements*: Generate concepts that meet customer requirements, evaluate current designs and superior technologies, and analyze manufacturability of the product.

9.2.1.3.1.4 Phase/Gate 4—Optimization This phase was used to determine the robustness of the process by examining possible problems and the effects of these problems. These potential problems will be reviewed to determine preemptive prevention procedures.

> *Checklist*: Design failure mode and effects analysis
>
> *Scorecard requirements*: Identify several potential failure modes, reduce risk priority number below a specified target.

9.2.1.3.1.5 Phase/Gate 5—Validate   This phase was used to take the project's final design to the customer who then reviewed the prototype. Other evaluations in the phase were conducted to test the manufacturability and the reliability of the product.

> *Checklist*: Customer review, manufacturing capability studies, mean time to failure study

The project plan is shown in Figure 9.2. It shows the tasks, task durations, start and end dates, predecessors of each task, and the resources required to perform each task. The project started January 5 and finished April 28.

9.2.2 Identify Phase Case Discussion

1. DFSS Project Charter: Review the project charter presented.
   a. A problem statement should include a view of what is going on in the business and when it is occurring. The project statement should provide data to quantify the problem. Does the problem statement provide a clear picture of the business problem? Rewrite the problem statement to improve it.
   b. The goal statement should describe the project team's objective and be quantifiable, if possible. Rewrite the goal statement to improve it.
   c. Did your project charter's scope differ from the example provided? How did you assess what was a reasonable scope for your project?

2. Project Plan
   a. Discuss how your team would develop their project plan and how they assigned resources to the tasks. How would the team determine estimated durations for the work activities?

9.3 Define Phase Activities

1. *Collect VOC*: Create a VOC survey to understand the current and potential customers' requirements.

Task	Duration	Start Date	End Date	Predecessors	Resources
Design for Six Sigma Jewelry Box Project Plan	78 days	1/5	4/22		
Phase/Gate 1 Identify	5 days	1/5	1/11		
Create project charter	2 days	1/5	1/6		
Perform stakeholder analysis	2 days	1/7	1/8	3	Team
Create project plan	1 day	1/11	1/11	4	Team
Phase/Gate 2 Define	38 days	1/5	2/25		
Gather Voice of the Customer	15 days	1/5	1/25		Team,Customer
Perform Market Availability Analysis	10 days	1/26	2/8	7	Team
Perform Competitive Benchmarking	5 days	2/9	2/15	8	Team
Perform Kano Analysis	5 days	2/16	2/22	9	Team
Perform Quality Function Deployment	3 days	2/23	2/25	10	Team
Phase/Gate 3 Design	17 days	2/26	3/22	11	
Generate concepts	10 days	2/26	3/11		Team
Use Pugh Concept Evaluation to select concepts	7 days	3/12	3/22	13	Team
Phase/Gate 4 Optimization	8 days	3/23	4/1	14	
Perform Design Failure Modes and Effects Analysis	8 days	3/23	4/1		Team
Phase/Gate 5 Validation	15 days	4/2	4/22	16	
Prepare prototype	15 days	4/2	4/22		Team
Review with customer	5 days	4/2	4/8		Team,Customer
Perform manufacturability study	10 days	4/2	4/15		Team
Perform mean time to failure analysis	10 days	4/2	4/15		Team

FIGURE 9.2
Project plan.

2. *Identify critical to satisfaction (CTS) measures and targets*: Based on the VOC, determine the CTS measures and then develop targets using benchmarking data.

3. *Translate VOC into technical requirements*: Using the CTS measures and targets, identify the technical requirements for the product.

9.3.1 Define Phase

9.3.1.1 Collect Voice of the Customer (VOC)

The voice of the customer refers to a responsive and innovative process that captures and relays the requirements of the customer/consumer in the form of feedback to the manufacturer/retailer. The purpose of the VOC is to provide the customer with the best-quality product and/or service in its class.

The team collected the voice of the customer using a survey. This survey consisted of nine simple questions, some of which also allowed the customers to make their own comments. The survey questions are shown in Figure 9.3.

Based on the availability and resources, our survey was distributed to college undergraduate and graduate women at the Missouri University of Science and Technology campus, Rolla, Missouri. The customers' more crucial needs of a traveling jewelry box were identified as follows:

- 68% prefer a traveling case that carries more than jewelry.
- 63% prefer brighter colors.
- 63% prefer lots of storage.
- 63% prefer to purchase such an item at a department store.
- 58% prefer a case that is compact and easy to clean.
- 58% prefer spending no more than $20 on such a case.

More information obtained from the survey suggested that 79% of the people who responded to the survey wanted a traveling jewelry box that carried makeup and other day-to-day personal items. Also, the survey pointed out that the most likely reasons a traveling jewelry box would be replaced were if it broke, became too dirty, or did not possess adequate storage.

Therefore, based on the survey responses, our team identified the basic customer needs for our product. They are as follows:

- Aesthetically pleasing design and colors
- Traveling case that can carry more than jewelry, such as makeup
- Lots of storage
- Compact

Dear Participant,

Thank you for taking to time to help our group gather data on what women want in a traveling jewelry box. The data that will be provided by you will be used to in our Design for Six Sigma project with goals to design a quality product that meets the customer's requirements in which we will translate into a final good. This final good will have several prototypes which will fulfill and possibly even exceed the customer's expectations. Your answers are completely anonymous and confidential, and will only be seen by the design team. If you have any questions, please contact any of the design team via the Missouri University of Science and Technology Miner mail address book.

Thank you!
Robert Jackson, Kayla Simmons, & Lyndon Chen

1. I usually/ will take traveling trips requiring packing jewelry (select one)?
    ___ less than once per year
    ___ 2-4 times per year
    ___ 5-8 times per year
    ___ 9 or more times per year

2. The type of traveling you participate in is (circle a number):

Mostly Business		Mix of Both		Mostly Leisure/Vacation
1	2	3	4	5

3. I prefer my traveling jewelry casing to have (select all that apply):
    ___ Neutral Colors (i.e. black, white, brown)
    ___ Bright Colors (i.e. lime green, yellow, pink)
    ___ Regal Colors (i.e. burgundy, dark blue, dark green)
    ___ Designer Patterns(i.e. Coach, Louis Vuitton, Gucci)

4. When traveling, I prefer to have a jewelry case that is/ has (select all that apply):

    ___ A lot of storage   ___ Hard Cased
    ___ Built in lock   ___ Interchangeable Colors
    ___ Carries more than Jewelry   ___ Jewelry Cleaning Kit
    ___ Compact Design   ___ Mirror
    ___ Drawers   ___ Removable compartment(s)
    ___ Easy to Clean   ___ Ring securer
    ___ Expandable   ___ Soft Cased
    ___ Flexible   ___ Stackable/ Removable trays
    ___ Other(s): _____
                  _____
                  _____

5. List the average number of items you carry when you travel.
    _____ Necklaces
    _____ Bracelets
    _____ Rings
    _____ Earrings
    _____ Watches
    _____ Other: _____

FIGURE 9.3
Survey.

- Easy to clean
- Available for purchase at a department store
- Inexpensive

The group studied these responses further to determine more specific customer needs, which are listed below:

- Compact outer shell
- Expandable compartments for lots of storage
- Robust design and less moving parts ensure reliability
- Compartments designed with an easy-to-clean material, such as plastic
- Affordable for college students
- Case appears attractive, perhaps with interchangeable shells to manipulate looks
- Case is safe for the user and for the items it holds
- See-through compartments or labeling capability
- Mirror for application
- Hard outer casing, but flexible to decrease damage

9.3.1.2 Identify Critical to Satisfaction (CTS) Measures and Targets

9.3.1.2.1 Competitive Benchmarking

In our research, the team discovered most store locations do not carry a specific traveling jewelry box. In most cases, customers are forced to purchase bulky caboodles or tackle boxes in order to safely separate their jewelry for travel. However, some Web sites do allow customers to purchase traveling jewelry accessories. The team's extensive research regarding traveling jewelry cases among market competitors is shown in Figure 9.4.

9.3.1.3 Translate Voice of the Customer (VOC) into Technical Requirements

9.3.1.3.1 Quality Function Deployment

Originating in Japan, quality function deployment (QFD) is a systematic process to integrate customer requirements into every aspect of the design and delivery of products and services. QFD is a collection of matrices used to facilitate group discussions and decision making. The QFD matrices are constructed using affinity diagrams, brainstorming, decision matrices, and tree diagrams.

QFD can lead to many great discoveries, such as

- Better understanding of customer requirements
- Increased customer satisfaction
- Reduced time to market and lower development costs
- Structured integration of competitive benchmarking into the design process
- Increased ability to create innovative design solutions
- Enhanced capability to identify those specific design aspects that have the greatest overall impact on customer satisfaction

Features	Competitor A	Competitor B	Competitor C	Design Team
Dimensions	9½" L × 4" W × 2" D	2.5" sq × 5" H	2 3/8" H × 5 1/2" W 7 1/2"D	8"H × 8"W × 8" D
Compact design	Yes	Yes	Yes	Yes
Lots of storage	No	No	Yes	Yes
Reliable	Yes	Yes	Yes	Yes
Location availability	Online or in store	Online exclusive	Online exclusive	Department stores
Price	$22 without shipping	$49 without shipping	$25 without shipping	$19.95 without shipping
Aesthetics	Only in silver	Only in pink or coffee	Only in red	Interchangeable color shells
Safe	Yes	Yes-user/ No-Cargo	Yes	Yes
Item accessibility	Easy	Average	Average	Easy
Bonus features	None	Mirror	Mirror	Mirror, Lock
Hard surfaced	Yes—rubber coated	Yes—leather covered plastic	No—padded faux suede	Yes
Damage resistant	Yes—Velvet antitarnish liner	No	No	Yes
Carries more than jewelry	No	No	No	Yes

FIGURE 9.4
Competitive benchmark.

- Better teamwork in cross-functional design teams
- Better documentation of key design solutions

QFD has two main approaches: the Four Phase Approach and the Matrix of Matrices, also known as the House of Quality (HOQ). Our group determined that the HOQ approach was the best fit for this project. The House of Quality is shown in Figure 9.5.

9.3.1.3.2 Discussion of House of Quality

Quality characteristics, also known as design requirements or "hows," were brainstormed and analyzed by the team in a similar fashion. Based on the customer-demanded requirements, our group developed possible ways to satisfy needs and avoid any adverse reaction to our product. These "hows" are the measurable implementations used to ensure all anticipated customer requirements are met. Quality characteristics and rankings were based on educational assumptions and information gathered from the competitive analysis. The demanded quality portion was validated, prioritized, and

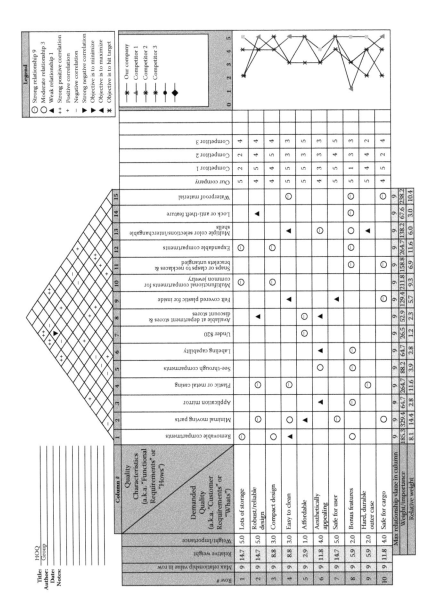

FIGURE 9.5
Quality function deployment House of Quality.

benchmarked. The weight/importance for each of the 10 customer requirements was based on the number of customers who felt a traveling jewelry case should provide these requirements. Each requirement relating to the surveys was tallied based on the number of times it was mentioned in Question 4 and Questions 6 through 9 of the survey. There were approximately 30 customers who responded to the survey. The customer responses include correlations between customer requirements as well. Figure 9.6 shows how the requirements were ranked. The rankings (weight/importance) can be found on the left-hand side of the House of Quality.

In the interrelation matrix (the center of the HOQ), our group compared each customer requirement to each quality characteristic and then determined the strength of each relationship and denoted that relationship with a symbol referenced in the HOQ's legend. In the event of a strong relationship, a "Θ" was implemented in the corresponding box.

A moderate relationship was denoted as an "O," and a weak relationship was denoted as "▲". If no relationship exists, then the corresponding box is left blank.

The "roof" matrix of the HOQ is used to identify whether the technical requirements that characterize the product support or hinder one another. As in the interrelationship section, our group considered the pairings of the technical requirements and for each pairing asked: "Does improving one requirement cause deterioration or improvement in the other technical requirement?" In the event of deterioration, the correlation between each requirement is considered a negative correlation and is further analyzed to determine the strength of deterioration. In the event of improvement, the

Customer Requirement	Ranking	Ranking Reasoning
Lots of storage	5	Survey responses
Robust/reliable design	5	Engineers' decision
Compact design	3	Survey responses
Easy to clean	3	Survey responses
Affordable	1	Survey responses
Aesthetically appealing	4	Survey responses
Safe for user	5	Engineers' decision
Bonus features	2	Survey responses
Hard, durable outer case	2	Survey responses
Safe for cargo	4	Engineers' decision

FIGURE 9.6
Customer requirement ranking.

correlation between each requirement is considered positive and is further analyzed to determine the strength of the improvement. Each correlation that exists is denoted with a symbol established in the HOQ's legend. If no correlation exists, the correlation cell is left blank.

In the competitive analysis, our group researched and compared other items in the market similar to our jewelry case. The rankings among the three competitors and our product were analyzed solely based on research, educated opinions, and customer feedback. Although the analysis shows that our product should compare above most of the competitors, our product has some room for improvements. For instance, our case is not as compact as some of the other cases; however, in order for the case to provide a lot of storage space it needs to have a complementary outer casing. Another example could be the case's safety features for the cargo. The inner case material is reliable and safe for common day-to-day items, but if the customer chooses something not expected, the material may not be the best fit. Next, our team prioritized the quality characteristics and the customer requirements with calculations that allowed the team to focus on the key requirements. Ultimately, by focusing on the requirements that had the greatest positive impact on the design, our group should be able to promote customer satisfaction as well as make our product more appealing to similar items in the market.

After the House of Quality was completed, our group was able to determine the importance of each quality characteristic. This ultimately helps determine which functional requirements our group should implement in order to increase customer satisfaction and product success. Based on the weight of importance, our group found that designing a case with minimal moving parts should be a priority for the safety of our users and reliability of our product. The next crucial requirement is having the primary casing material be either metal or plastic for durability and the ability to expand for the desired amount of storage. The requirements should be prioritized by the relative weight determined at the bottom of the HOQ. The requirement with the highest relative weight is the most critical, just as the requirement with the lowest rating should be the least critical. Our group will simply execute the functional requirements in descending order of relative weight. This is essentially how our group determined the product concepts.

9.3.1.4 Kano Analysis

The Kano model of customer satisfaction classifies product attributes based on how they are perceived by a customer and their effect on the customer's satisfaction. These attributes are divided into three categories: basic needs, linear satisfier, and delighters. Basic needs include attributes that the customer specifically requests. Linear satisfiers include attributes that are not specifically requested but are expected to be present. Delighters are attributes that are unknown to the customer but are pleasing to them once they are discovered. The categories are shown in Figure 9.7.

Category	If Present	If Not Present
Basic needs	Satisfied	Dissatisfied
Linear satisfier	Neither satisfied nor dissatisfied	Very dissatisfied
Delighter	Very satisfied	Neither satisfied nor dissatisfied

FIGURE 9.7
Kano model categories.

The Kano model for this product is shown in Figure 9.8. The delighters are the lock option, changeable style, storage options, and being personalizable. The satisfiers are able to carry more than jewelry, have a mirror, be cleanable, and be reasonably priced. The basic needs are that the jewelry box is durable, secure for travel, easy to transport, easy to use, and has online support for the product.

9.3.2 Define Phase Case Discussion

1. How did your team perform the VOC collection? How could the VOC collection be improved?
2. Did your team create and distribute a customer survey, and if so, what is the appropriate statistical analysis to perform to identify the importance of the customers' requirements?
3. Did you perform a quality function deployment? How did you identify the technical requirements and the correlations between customer and technical requirements?
4. What is the value of using the Kano model in your VOC analysis?

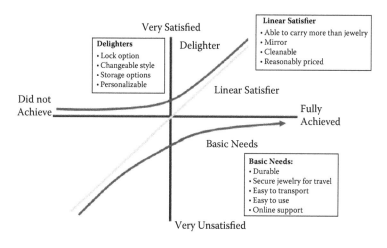

FIGURE 9.8
Kano model for product.

9.4 Design Phase Activities

1. *Identify Design Concepts*: Develop the design concepts.
2. *Design the Product and Select the Concepts*

9.4.1 Design Phase

9.4.1.1 Identify Design Concepts

This phase was crucial to the entire project. The development of concepts was looked at in three ways: aesthetics, functionality, and cost constraints. With the collaboration of several different viewpoints, concepts were developed and redeveloped to meet customer satisfaction.

9.4.1.2 Concept Generation

Our concept generations were designed based solely on the customer's requirements from the voice of the customer. The designs were then examined to identify which were able to be manufactured at a reasonable price (compared to the retail price) and which would go beyond satisfying as many customers needs and wants as possible.

Concept A *(Expanding Jewelry Rack)*: Concept A is a box with several compartments including four miscellaneous compartments, a bracelet roll, and four rolls for other jewelry. It also provides an expanding mechanism that allows for necklaces to hang vertically when the case is open and allows for storage of the other jewelry below the scissor lift. The lift would be able to retract back into the edges of the case if necklaces were not needed and would lie back down in the middle compartments. The case would need to be laid flat for the lift to operate properly, otherwise the necklaces may tangle. Concept A is shown in Figure 9.9.

Concept B *(Fold-Down Jewelry Case)*: Concept B is similar to Concept C, but it is much smaller in scale. It has a top compartment with hooks to hang the necklaces and place them into the pouch. It features a ring holder at the base and several compartments with a latched cover to secure them in place. This product would be a hard plastic or metal with velvet interior fabric and a cushioned leather exterior. Concept B is shown in Figure 9.10.

Concept C *(Fold-out Flat Jewelry Case With Rearrangeable Compartments)*: Concept C would provide a fold-out layout case that would be laid flat for full viewing and access with editable walls and straps. There are several different wall sizes and other jewelry storage items that can be interchanged using powerful magnets in the casing.

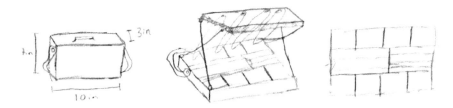

FIGURE 9.9
Concept A.

This concept has two Plexiglass® covers that can hold all items in place and allow the customer to easily see what he or she has. The item can support several items besides jewelry. Concept C is shown in Figure 9.11.

Concept D (Hard-Shell Jewelry Box With Removable Tray): Concept D is a hard casing made out of antitarnish tin-coated steel. With a fastener to ensure the case securely shuts (possible key entry), this item is composed of a set of removable trays made out of felt-covered plastic allowing for easy cleanup. The two layers allow for different items to store: the top for jewelry and the bottom for makeup and other toiletry items. Concept D is shown in Figure 9.12.

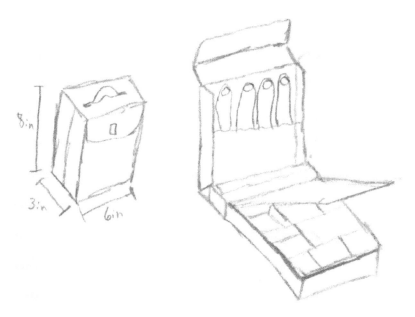

FIGURE 9.10
Concept B.

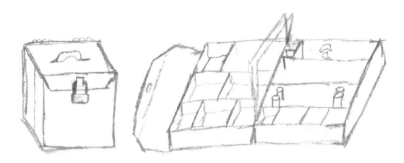

FIGURE 9.11
Concept C.

Concept E (Vertical All Hanging Jewelry Case): Concept E is a flexible roll up and out design for easy transport whether in luggage or just carrying by hand. The back has several clear compartments and necklace and bracelet snaps to ensure all items stay secure and untangled. The lining would provide compartmental pouches that would be accessible when rolled out and hung up but be storable when rolled up into the two semihard circular casings at the bottom of the roll-out sheet. When rolled up to be stored away the case would resemble a cylindrical tube with carrying straps on the exterior shell. Concept E is shown in Figure 9.13.

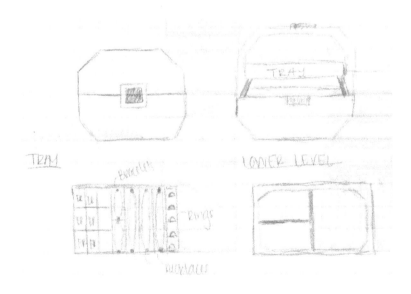

FIGURE 9.12
Concept D.

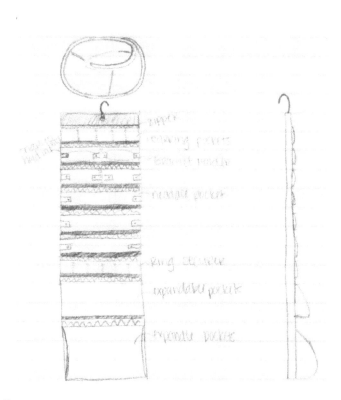

FIGURE 9.13
Concept E.

Concept F (Stackable Jewelry Tray Box): Concept F provides stackable trays that snap together and would be secured, allowing for a customizable size depending on the needs of the customer for that trip. Additional needs could be catered to by adding another tray to the case if, for example, the customer wanted to bring more jewelry or have a separate compartment for toiletries. The casing would be a hard shell with a rubberized coating. A mirror could be recessed into the case top, and a replacement tray or top could be purchased if the customer wanted an updated color or style or if damaged without needing to repurchase the entire jewelry box. The price could be variable with a smaller basic starter box with optional feature trays being bought later on or incorporated in more expensive models. Compartments would not be rearrangeable in each tray, but material linings would be different for each type of tray used; for example, a felt lining would be used for jewelry trays while a rubber lining would be used for a makeup tray for easier cleaning after any spills. New trays could be purchased if the customer wants to be fashionable and change the shell coloring or design or even mix colors for a rainbow effect. Concept F is shown in Figure 9.14.

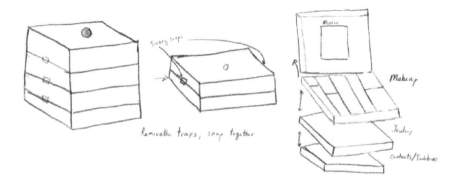

FIGURE 9.14
Concept F.

Concept G (Clamshell Case with Two Separate Compartment Halves):
Concept G provides a compact clamshell casing that would have pegs on one half to organize jewelry and compartments on the other side of the clamshell for storing more jewelry or makeup. This slimmer case would allow for easier portability at the expense of space. The sleek profile would be more stylish than older box-style cases and be made of a hard plastic shell. Concept G is shown in Figure 9.15.

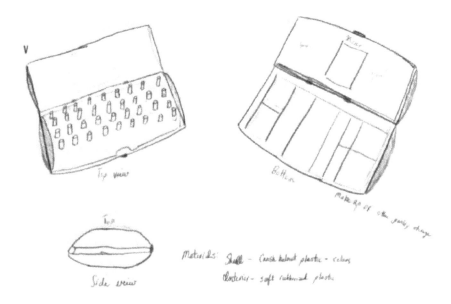

FIGURE 9.15
Concept G.

9.4.1.3 Design the Product: Select the Concept

9.4.1.3.1 Pugh's Concept Selection

Our team created sketches to prove concepts of storing items in a jewelry case and the possible results that method would form. The team developed seven concepts, although a few are similar enough to have shorter explanations. A Pugh's Concept Selection matrix was developed to narrow the field on possible designs for our final product. Comparison was performed against a product from the HOQ benchmarking section. Although these concepts are the focus, certain aspects of the grading were partly evaluated by how the concept would result in a finished product, and the sketches help illustrate the point. The Pugh's Concept Selection matrix is shown in Figure 9.16.

Out of the concepts compared, several did markedly better than others, especially when taking into account the importance of the selected criteria. The Stackable Jewelry Box showed the most promise in the score matrix and was picked to be the concept to design for our final product. The other concepts that scored well, such as the hard-shell jewelry box and the clamshell case scored high for similar reasons; therefore, the team decided not to combine them for a better product. Concept C, however, scored well with its rearrangeable magnetic compartment that is maybe something we keep on the back burner to use if we see the availability to use it within the stackable tray box design. The selected concept will be incorporated in a jewelry case similar to its sketch and will have the following features:

- The dimensions of the box will be a square 8" × 8".
- The variable height depends on the number of trays used but will start at a height of 8" that consists of a top cover of 1" height, two 2" trays (makeup and jewelry), and one 3" (toiletries/miscellaneous storage) in the base model.
- The trays will be tailored for individual purposes.
  - A tray designed for makeup will have a rubber inner lining that is easily cleanable from any spilled powders or liquids.
  - The trays for jewelry will be lined with felt and cushioning to handle the delicate items.
  - A toiletries tray would have compartments for items such as a toothbrush, clippers, contact solution, and so forth.
  - An electronics tray would have special slots to charge stored items such as an iPod/MP3 with a cord that would connect to the back of the box for easy charging. The specialized tray would have built-in speakers to make it a musical jewelry box. This option would be an aftermarket add-on and would not be included in the initial purchase for cost pricing.

Criteria	Competitor A	Weight	Concept A	Concept B	Concept C	Concept D	Concept E	Concept F	Concept G	Datum[a]
Compact design	4	3	S	S	–	–	–	+	+	
Lots of storage	1	5	S	S	+	S	+	+	+	
Robust/reliable design	3	5	–	S	–	+	–	+	S	
Affordable/cheap	5	1	–	S	–	S	S	–	–	
Aesthetically appealing	3	4	+	S	+	+	–	+	S	
Safe for users and cargo	3	5	–	S	S	+	–	+	S	
Bonus features	2	2	S	S	+	S	+	+	+	
Hard surface casing	5	2	S	S	S	S	–	S	S	
Damage resistant	4	5	–	–	–	S	–	S	S	
Capability of carrying more than jewelry	1	5	S	+	+	+	+	+	+	
Sum of (+)			1	1	4	4	3	7	4	
Sum of (–)			4	1	4	1	6	1	1	
Sum of (s)			5	8	2	5	1	2	5	
Net score = sum of (+) – sum (–)	s		–3	0	0	3	–3	6	3	
Weighted score rank (weight × value in column, where + = 1, – = –1 and s = 0)			–12	0	2	16	–12	28	14	
			6	5	4	2	6	1	3	

[a] Datum is Competitor A from market analysis survey.

FIGURE 9.16

Pugh Concept Selection matrix.

- Trays will stack upon each other and will seal to one another but have a button-latch system that will guarantee no trays will accidentally separate during transportation or rough use.
- The outer shell will be made of a hard plastic that is coated in a rubberized finish.
- The exterior will be offered in a variety of colors and designs that can be mixed and matched for fashion appeal.
- A kids' version could be offered with colorful characters that could be swapped around in order to create humorous patterns or effects.

9.4.1.4 Design for X Methods

The product development stage has been carefully examined by the designers and the manufacturing engineers to ensure that the final design not only meets the customer's criteria but also is reasonable for a manufacturer to produce it. This area along with others will be designated as the Design for X method. There are several areas in this method (those underscored will be examined for this product):

- Design for produceability (manufacturability)
- Design for assembly
- Design for reliability
- Design for serviceability (maintainable)
- Design for environment
- Design for testability (measurable functions)

In this section, our team examined several areas of what items will need to be produced and what items can be outsourced. Figure 9.17 includes a list of the predicted parts needed to produce a single product. As time progresses, many of the same operations will be used to innovate and continue progression toward customer satisfaction. Meetings with suppliers and production will be needed to ensure the success of the manufacturing of this product.

9.4.1.5 Design for Assembleability

When the design for manufacturability is examined, design for assembleability should be aligned with it. This section will look at the product and its subsystems to determine the best way to assemble it and reduce variability. In the design of this product, our team wanted to focus on keeping the product design simple and reducing part count to minimize assembly time.

Item Number	Part Name	Method of Manufacture	Produced In-House	Materials Needed	Tools/Entities Needed	Labor Needed
1	Trays	Polymer injection molding	Yes	Polymers billets	Injection molding machine CNC for molds	Maintenance machinist
2	Latch	Standard part	No	n/a	Outside vendor adhesive/riveter	Operator
3	Buttons	Polymer injection molding	Yes	Polymers billets	Injection molding machine for molds	Maintenance machinist
4	Springs	Standard part	No	n/a	Outside vendor	Operator
5	Coatings	Powder coating rubberizing	Yes	Coating rubber materials TBD	Air brush dipping vat Spray booths Ovens Racks	Operator
6	Fabrication	Hand stitching	Yes	Leather thread adhesive	Sewing machines Fabric cutters	Operator
7	Metal clasp	Standard part	No	n/a	Outside vendor adhesive/riveter	Operator
8	Rubber seals	Standard part	No	n/a	Outside vendor Adhesive	Operator
9	Adhesives	Standard part	No	n/a	Outside vendor	Operator

FIGURE 9.17
Design for manufacturability.

Item Number	Operation	Number of Components	Stand-Alone Station?	Station Objectives	Approximate Time per Part
1	Tray formation	1	Yes	Form tray from injection machine and allow to cure	10 min
2	Attach latch	1	No	Attach latch to top	20 sec
3	Attach rubber seals	1	No	Adhere seal to tray	45 sec
4	Sew handles	1	No	Wrap handle with leather and sew	10 sec
5	Powder coat	1 or more	Yes	Coat trays with material	30 min
6	Button assembly	4	Yes	Assemble button release	20 sec
7	Attach handles	1	No	Attach handle to lid	20 sec

FIGURE 9.18
Design for assembleability.

The item will first be hand built, then as the product becomes more widely accepted, we will switch to a more automated process. These elements are shown in Figure 9.18.

9.4.2 Design Phase Case Discussion

1. How did you generate your design concepts?
2. How did you determine how your concepts compared using the Pugh Concept Selection matrix?
3. How did you derive the best combination of your design elements from each concept?

9.5 Optimize Phase Exercises

1. Implement pilot process.
2. Assess process capabilities.
3. Optimize design.

9.5.1 Optimize Phase

In this phase, the team reviewed the selected superior concept to identify possible flaws and their effects and probable solutions on how to reduce them. The team also compared the influence of one factor on another and outside noise factors.

9.5.1.1 Process Failure Mode and Effects Analysis

A Process Failure Mode and Effects Analysis (PFMEA) was used to answer the question of "what if." This system is in the design phase so an analysis to determine worst-case scenarios with failures of components can be used to adjust the design to turn the variables into constants. Our team has gathered much of the data for the PFMEA based on what we know about similar systems and how they function and problems that may occur to similar components in the systems. We based the risk priority number (RPN) on our previous experience. Using the RPN number, the team can prioritize problem areas and make changes as needed. The PFMEA is shown in Figure 9.19.

This tool was useful for identifying problems that may occur during the manufacturing and assembly of this product. This tool can be modified and reevaluated when production is online and there is a better idea of problem areas.

9.5.2 Optimize Phase Case Discussion

1. How did you define the design of experiment design to use?
2. How did you determine which factors and levels were significant to your design?
3. How did you determine the appropriate number of replications for your experiment?

9.6 Validate Phase Activities

1. Validate process.
2. Assess performance, failure modes, and risks.
3. Iterate design and finalize.

9.6.1 Validate Phase

Modeling of variation sensitivities across the integrated technology platform and verification of Concept F and the resulting product are necessary to ensure that our design meets the customers' needs. First, a customer survey

Item Number	Item	Function	Potential Failure Mode (Possible Failure)	Potential Failure Effects	Severity	Potential Causes/ Mechanisms of Failure	Occurrence	Recommended Actions	Detection	Risk Priority Number (RPN)
1	Tray molds	Used in the mold injection	Trays do not mold clearly	Trays will have voids	3	Improperly trained staff	4	Training and test out	4	48
2	Velvet/fabric wrap	Hand-wrapped on trays	Does not adhere well	Rips and causes rework	2	Improperly trained staff	4	Training and test out	8	64
3	Coatings	Sprayed on in booth	Uneven coating	Chipping or voids	2	Uneven spray	6	Automate for standardization, Sample QA	8	96
4	Snap loops	Sewn into place	Positioned correctly	Low-quality product/rework/ strength problems	3	Unevenly stitched	7	Have fixture with Poke Yoke, strength testing	2	42
5	Button release	Hand installed into place	Buttons are not secured	Broken buttons/ springs	4	Incorrect installation	3	QA 100%	7	84
6	Lid latch	Adhered/sewn into place	Positioned correctly	Low-quality product/ rework/strength problems	2	Unevenly stitched	7	Have fixture with Poke Yoke, strength testing	2	28
7	Handle	Sewn into place	Positioned correctly	Low-quality product/rework/ strength problems	2	Unevenly stitched	7	Have fixture with Poke Yoke, strength testing	2	28

FIGURE 9.19
Process failure mode and effects analysis.

Survey Results Percentage by Category			
Question	Strongly Disagree/ Disagree (%)	Neutral (%)	Agree/ Strongly Agree (%)
1. Does this product meet the ability to offer a satisfactory amount of storage space for your jewelry, makeup, or other wanted products?	10	10	81
2. Do you feel like this product is durable and stain resistant?	5	14	81
3. Is this product a compact solution you would be willing to travel with?	10	29	62
4. Is this product aesthetically pleasing or fashionable?	0	5	95
5. How interested would you be if we could offer you this product for under $60?	10	5	86
6. Would you appreciate a line of aftermarket accessories/parts like different designs, exterior case material, replaceable trays, or the musical jewelry box add-on?	0	14	86

FIGURE 9.20
Survey responses.

was conducted with a picture, description, and small model to 21 collegiate women. The women surveyed were queried about how they felt our product met their needs using a scale of (1 to 5) for six questions using a Likert scale of 1 = strongly disagree, 3 = neutral, and 5 = strongly agree. The survey results are shown in Figure 9.20.

Further verification will need to be performed over the manufacturing of this product and the facilities' ability to adhere to our design requirements. This step will have to be completed at a later date due to capital cost. However, a prototype was fabricated for display. The results are encouraging enough that we may have come close to our customers' needs.

9.6.2 Validate Phase Case Discussion

1. How did you identify potential failure modes of your product?
2. How did you identify potential risks of your product?
3. How would you assess the potential market for your product?

9.7 Conclusions

Working as a team, we were able to use the DFSS methodology to redesign the way women can carry their jewelry and other accessories to a better degree of satisfaction. We were able to gather their needs and create a product that is variable in size and durable in nature while remaining attractive and organized. The ability for further business is designed into the product for later purchases for customization or feature. DFSS also allowed for consideration to the manufacturing of the product and what specifications were needed to fulfill the customer requirements in the given boundaries set up by the team. While this is a lengthy process to go through, the benefit of providing a higher-quality product by design is tangible. We are prepared with the Design for Six Sigma methodology to satisfy the goals of our project and our customer requirements.

10

Design of an Optical Mouse—A Design for Six Sigma Case Study

**Priya Dhuwalia, Soundararajan Chandra Moleeswaran,
Swathi Priya Pedavalli, Vishwanath Narayanan,
Elizabeth Cudney, and Sandra Furterer**

CONTENTS

10.1 Project Overview

The design of the existing optical computer mouse is redundant, and the aim of this project is to develop new designs implementing Design for Six Sigma (DFSS) principles for improving the existing characteristics to arrive at a final model that is better in aesthetic value, design, and compatibility. In addition, ergonomics will be taken into account to arrive at a better-quality product. Design for manufacture (DFM) principles will also be implemented, thereby eliminating unwanted fasteners and parts for assembly. This will lead to the reduction in material and assembly costs and thereby decrease the overall cost of manufacture that leads to improved profit levels. This product will cater to the needs of customers in the medium to high budget range.

The goal of the project is to reduce manufacturing cost by 10%, increase market share by 5%, and design an optical mouse with better aesthetic look, design, and compatibility. Reducing the number of fasteners will result in savings for the customers and also a sophisticated design. The data were collected from the potential customers of the product by distributing a survey to different categories of people. According to the data obtained and analyzed, the voice of the customer (VOC) was implemented in the product. The team also focused on designing the product to reach its maximum capacity and functions using DFSS principles.

The objectives of the project are to (1) apply the DFSS concepts to modify the design of an existing optical computer mouse; (2) improve the existing characteristics to arrive at a final model that is better in aesthetic value, design, and compatibility; (3) improve the ergonomics of the product; and (4) implement design for X (DFX) principles and reduce the material and assembly costs.

This project will focus on developing an optical mouse. Concepts and designs generated will be evaluated in order to determine the best design and create a model to demonstrate its usefulness. In order to optimize the model, experiments will be discussed but not conducted due to time and resource constraints.

10.2 Identify Phase

10.2.1 Identify Phase Activities

It is recommended that students work in project teams of three to four students throughout the DFSS case study.

1. *Develop Project Charter*: Use the information provided in the Project Overview section to develop a project charter for the DFSS project.
2. *Team Ground Rules and Roles*: Develop the project team's ground rules and team members' roles.
3. *Develop Project Plan*: Develop your team's project plan for the DFSS project.

10.2.2 Identify

10.2.2.1 Project Charter

The first step was to develop a project charter.

Project Name: Design of an optical computer mouse

Project Overview: The objective of the project is to apply the DFSS concepts to modify the design of an existing optical computer mouse.

Problem Statement: The design of the existing optical computer mouse is redundant; therefore, the aim of this project is to develop new designs by implementing DFSS principles for improving the existing characteristics to arrive at a final model that is better in aesthetic value, design, and compatibility. Ergonomics will be taken into account to arrive at a better-quality product. Design for manufacture (DFM) principles will also be implemented, thereby eliminating unwanted fasteners and parts for assembly. This will lead to the reduction in material and assembly costs and thereby decrease the overall cost of manufacture, leading to improved profit levels.

Customer/Stakeholders: All computer users

Goal of the Project: Reduction in manufacturing cost by 10%

Scope Statement: This product will cater to the needs of customers in the medium to high budget range.

Projected Financial Benefit(s): Increase market share by 5%

10.2.2.2 Team Ground Rules and Roles

The team informally developed several ground rules for the project:

- Everyone is responsible for the success of the project.
- Listen to everyone's ideas.
- Treat everyone with respect.
- Contribute fully and actively participate.
- Be on time and prepared for meetings.
- Make decisions by consensus.
- Keep an open mind and appreciate other points of view.
- Communicate openly.
- Share your knowledge, experience, and time.
- Identify a backup resource to complete tasks when not available.

10.2.2.3 Project Plan

Once the idea was established (invent), a customer survey was developed and deployed in order to receive the specifications via the voice of the customer. The Gantt chart shows the phases of the project from start to finish, noting that the phases were not completed in parallel but instead in series; however, some of the phase subtasks were able to be completed in parallel. The phases of the project were *Identify, Define, Design, Optimize,* and *Validate* (IDDOV). Checklists were used as gates to each of the phases, and scorecards were maintained. During the *Optimize* and *Validate* phases, the scorecards were useful in determining which phases could be revisited in order to further optimize the IDDOV process.

IDDOV uses a phase/gate approach to develop the technology associated with a particular product development. It is divided into five phases. At the end of each phase the team reached a gate where the deliverables were checked to account for the achievements in the respective phases. Various checklists and scorecards can be built based on PERT or Gantt charts. A Gantt chart was developed to specify the time line for each task and ensure that the current tasks were being carried out in accordance with the planned schedule as shown in Figure 10.1.

10.2.3 Identify Phase Case Discussion

1. DFSS Project Charter: Review the project charter presented.

    a. A problem statement should include a view of what is going on in the business and when it is occurring. The project statement should provide data to quantify the problem. Does the problem statement provide a clear picture of the business problem? Rewrite the problem statement to improve it.

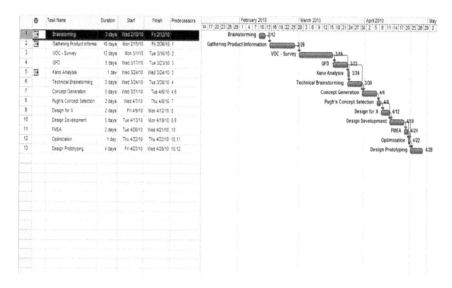

FIGURE 10.1
Project Gantt chart.

b. The goal statement should describe the project team's objective and be quantifiable, if possible. Rewrite the goal statement to improve it.

c. Did your project charter's scope differ from the example provided? How did you assess what was a reasonable scope for your project?

2. Project Plan

a. Discuss how your team developed their project plan and how they assigned resources to the tasks. How did the team determine estimated durations for the work activities?

10.3 Define Phase

10.3.1 Define Phase Activities

1. *Collect VOC*: Create a VOC survey to understand the current and potential customers' requirements.

2. *Identify critical to satisfaction (CTS) measures and targets*: Based on the VOC, determine the CTS measures and then develop targets using benchmarking data.

3. *Translate VOC into technical requirements*: Using the CTS measures and targets, identify the technical requirements for the product.

10.3.2 Define

The team started the commercialization process mainly by obtaining

1. The voice of the customer
2. The voice of technology

The voice of the customer (VOC) has to be utilized to develop the product in line with customer needs and requirements. To define the product line strategies and family plans, additional VOC can be obtained to target various market segments. The product may need modifications or additional features depending on the market segment under target. Using VOC and voice of technology (VOT), technology houses of quality can be developed to build a foundation on which the entire plan can be developed. The technology House of Quality can be used to benchmark criticality and priorities in technology needs. Both quantitative as well as qualitative approaches were carried out in order to elicit the VOC. An online survey was the primary source for quantitative analysis, while conversations with potential customers were the source for qualitative analysis.

> *Checklist*: Market segment analysis by interviewing customers and performing a market trend forecast to address the question, "Is there something like this out on the market yet?" In addition, benchmarking, gathering voice of the customer through a survey, affinity diagrams, quality function deployment (QFD), and Kano diagrams were utilized to ensure the team was heading in the right direction.
>
> *Scorecard requirements*:

- Studying the market forecast
- Benchmarking our product to others available in the market
- Gathering information on voice of the customer through an online survey
- Translating customer needs to useful metrics and ranking them through building our House of Quality
- Prioritizing customer requirements based on survey results
- Performing Kano analysis
- Completing the House of Quality

10.3.2.1 Identifying "Voice of the Customer"

In order to identify the needs of the customer, a survey consisting of essential questions was created. This survey was circulated among various communities. The team received 58 survey responses from the participants with a variety of backgrounds of which half are students and the other half employed.

10.3.2.2 Structuring and Ranking Customer Needs

According to the responses from the participants, these were the expectations and preferences:

57.1% preferred wireless

72.7% preferred just about the size of the palm

75.4% preferred a two-button clicker

66.1% were ready to invest <$20 for additional features

The other set of information obtained from the survey was that 73.6% expected more comfort, 52.8% design/aesthetics, 47.2% size, and 45.3% size of the mouse while buying one. Further key statistics include the following:

- *Type*: 29% prefer a wired mouse, 57.9% wireless, and the rest do not want to use a mouse
- *Size*: 1.8% bigger than the palm, 25% smaller, and 73.2% just about the size of the palm
- *Clicker*: 24% single clicker and 76% double clicker
- The majority (over 60%) of the participants carry the mouse less than 10% of the time they travel.
- 67% of the participants experienced discomfort while using the mouse for long hours.

Hence, based on these results the team identified the basic customer needs for our product as follows:

- Wireless
- About the size of the palm
- Two-button clicker
- Comfort
- Aesthetic appeal
- Economical

The responses given by the customers are given in Figure 10.2.

A majority of the respondents carry the mouse less than 10% of the time when traveling. In addition, 67% of the participants complained of experiencing discomfort while using the mouse for long hours. Hence, the team also concentrated on the aesthetics and comfort level while using the mouse.

Based on the requirements and the voice of the customer, the team rated the specified features and their significance in the design (see Figure 10.3).

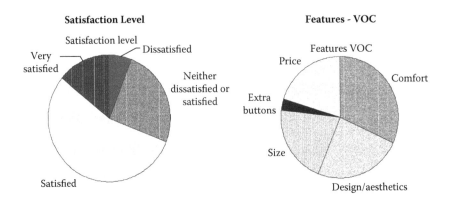

FIGURE 10.2
Customer survey results.

10.3.2.3 Analysis of Competitors in the Market

The leading competitors in the market were determined to compare against our product, and the features for their products were ranked based upon the Web research and face-to-face interactions with the current customers. Then a competitor benchmarking was performed by comparing the proposed product with those five competitors. The ratings given for customer-specified features for a specific brand are listed in Figure 10.4.

From the competitor benchmarking it was observed that the highest rating was given to Mouse D. Hence, it was considered as our datum mouse.

10.3.2.4 Quality Function Deployment

The next step was obtaining, translating, and deploying the voice of the customer into various phases of technology development and ensuring

Number	Feature	Importance
1	Type of mouse	2
2	Size of the mouse	4
3	Clicker	3
4	Scroller	4
5	Comfort level	5
6	Price	5
7	Portability	4
8	Sensitivity of movement	3

FIGURE 10.3
Customer needs ranking.

Feature	Mouse A	Mouse B	Mouse C	Mouse D	Mouse E
Type of mouse	3	5	3	5	3
Size of the mouse	4	3	2	5	2
Clicker	3	3	3	5	3
Scroller	3	5	4	5	3
Comfort level	3	4	2	5	2
Price	4	2	3	2	2
Portability	3	2	2	4	2
Sensitivity of movement	3	4	3	5	3

FIGURE 10.4
Competitor benchmarking analysis.

commercialization of the process during product design. The team defined the results in three categories: new needs, unique needs, and difficult to fulfill needs. From this information the team created the House of Quality.

The process known as quality function deployment (QFD) is commonly used to help further refine the VOC. Data from the VOC are mapped using the KJ model to know the requirements of the customer and determine the measures of performance. QFD produces a two-dimensional matrix called House of Quality. HOQ is a system used to transform data from the VOC into technical and performance requirements for the new product.

Using the voice of the customer the team built the House of Quality (see Figure 10.5). These are the steps the team used to build the HOQ:

Step 1: List the customer requirements.

Step 2: List the technical descriptors (characteristics that will affect more than one of the customer requirements, in development or production).

Step 3: Compare the two (customer requirements to technical descriptors) and determine relationships.

Step 4: Develop the positive and negative interrelated attributes and identify "trade-offs."

The team analyzed the present market to find probable competitors who are working on or possess the current technology the team planned to implement in the design. When this was done, the key potential competitors were Logitech and Apple. There is a separate section in the HOQ where the tradeoffs are marked and documented. This section in the HOQ helped the team with the initial identification of DFSS requirements. The competitive benchmarking showed the team what the customer expected that was not met by the current competitors, thereby surpassing them with the proposed design.

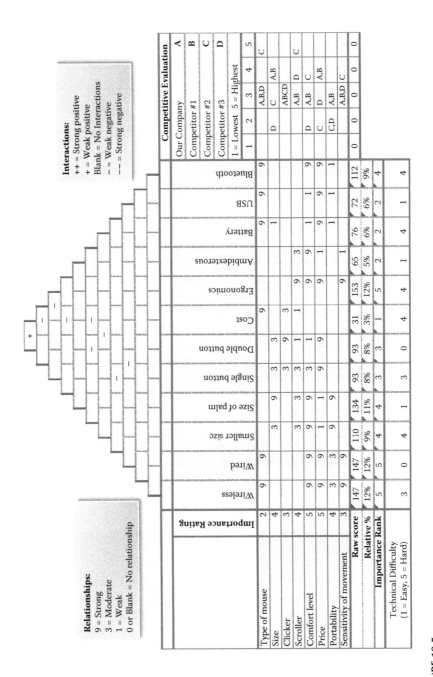

FIGURE 10.5
House of Quality.

10.3.2.5 Kano Analysis

When conducting a Kano model the team must focus on customer satisfaction while fulfilling basic needs of the product and expected quality. The successfulness of the product was determined by the results obtained from the Kano customer satisfaction survey. The delighter that excited most of the customers was the avoidance of discomfort by the current design when the mouse is extensively used for long hours. The survey yielded positive results of customer's investment for additional features. However, the model has been designed to provide more additional features for the same price, another delighter. The option of easy assembly and disassembly of the product excited the customer and would be a strong selling point, also enabling the customer to gain functionality of the product. The other delighters that the team considered and that resulted from the survey were the usage compatibility of the optical mouse on all possible surfaces and the single button with a double clicker. Figure 10.6 shows the Kano analysis.

10.3.3 Define Phase Case Discussion

1. How did your team perform the VOC collection? How could VOC collection be improved?
2. Did your team create and distribute a customer survey, and if so, what is the appropriate statistical analysis to perform to identify the importance of the customers' requirements?
3. Did you perform a QFD? How did you identify the technical requirements and the correlations between customer and technical requirements?
4. What is the value of using the Kano model in your VOC analysis?

Questions	Delighted	Normal	Expected
If the type of the mouse is wireless?			X
Size of the mouse about the size of the palm?		X	
Avoids discomfort when used for long hours?	X		
If the aesthetics of the mouse look better?			X
If comfort levels of operating increased?			X
If the cost of the mouse is maintained the same for additional features?	X		
If it is made compatible to use with all surfaces?	X		
Easy to assemble/disassemble?	X		

FIGURE 10.6
Kano analysis.

10.4 Design Phase

10.4.1 Design Phase Activities

1. Identify process elements.
2. Design process.
3. Identify potential risks and inefficiencies.

10.4.2 Design

This is the main phase of the technology development and modeling. Concepts have to be evaluated to determine if they satisfy all key requirements based on the technological concepts. Different design concepts in accordance with the customer requirements were developed in this phase and evaluated.

> *Checklist*: Concept generation technique (Theory of Inventive Problem Solving [TRIZ], brainstorming), design for manufacture and assembly, concept generation, affinity diagrams, Pugh concept evaluation and selection

> *Scorecard requirements*:

> - Generating seven design concepts that meet customer requirements
> - Evaluating superior concepts and superior technology to beat the market
> - Analyzing and studying feasibility of superior concepts—looking at the competitors
> - Adding value to the design by thinking "outside the box"

> *Concept 1—Wired to Wireless* (see Figure 10.7): A radio-frequency (RF) transmitter will be integrated in the mouse which will record the mouse movements and clicks, and then send them as RF signals to

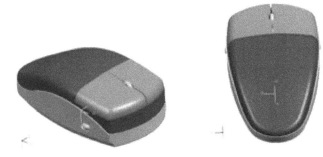

FIGURE 10.7
Design Concept 1.

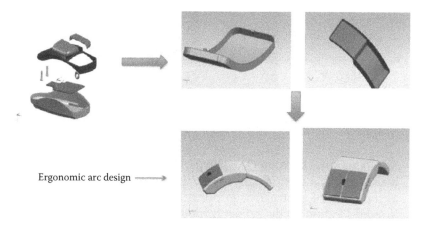

Ergonomic arc design ──────►

FIGURE 10.8
Design Concept 2.

the RF receiver. The RF receiver receives these RF signals, decodes them, and then sends these signals directly to the computer. 802.11b or 802.11g radio frequencies operating at 2.4 GHz will be used for the data transmission. Automatic pairing of the receiver and transmitter will take place, thereby eliminating the interference issue.

Concept 2—Elimination of Fasteners, Ergonomics (see Figure 10.8): In this concept, fasteners were substituted with hinges to reduce the cost of manufacturing, assembly time, and ease of design. Also, the design is modified to an ergonomic arc design, thereby decreasing the material required for manufacturing and improving the aesthetics of the product. Also, the reason for changing the design to an arc design is to eliminate carpal tunnel syndrome caused due to usage of the mouse for extended hours. This syndrome creates pain in the index and middle fingers and in the wrist because of the way of handling the mouse. At a resting position the human palm rests in an arc position, and the arc design fits well under the palm of the user eliminating this syndrome, hence it was taken into consideration.

Concept 3—Monoclicker Concept (see Figure 10.9): The clicker design has been modified such that there is a monoclicker now rather than two separate clickers for the left and the right. There has been a pivotal support that has been provided for the center of the mono-clicker to rest on, thus eliminating the time required to assemble the second clicker.

Concept 4—Ball Scroller (see Figure 10.10): Here the team introduced the concept of a "trackball" instead of a regular scroller. A trackball con-sists of a ball that is supported within the housing and is permitted

FIGURE 10.9
Design Concept 3.

to rotate in the housing. The ball protrudes through the top of the housing so that the user can supply motive force directly to the ball at the point of protrusion. In operation, the cursor is moved to a desired location on the screen through movement of a rollerball by the user, and a specific function is then selected by pressing one or more keys on the control device. Rotation of the ball by the user is transferred into two orthogonal directions such as X and Y coordinate directions to move the pointer on the computer screen in the X and Y directions.

This ergonomically friendly mouse is shaped to help avoid strain on the hands and fingers and help prevent carpal tunnel syndrome and other repetitive strain injuries. Trackballs have been proven to relieve wrist and hand strain that leads to carpal tunnel syndrome.

Concept 5—Laser Mouse (see Figure 10.11): After introducing the trackball and proposing designs that concentrated on aesthetics, the focus was on improving efficiency and precision of working, thereby introducing the concept of a laser. This concept uses a laser diode instead of a light-emitting diode (LED) to reflect the light off a surface, and then compare the images taken by the CMOS sensor, to determine the displacement from the original position. The coherent

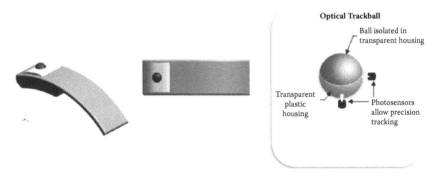

FIGURE 10.10
Design Concept 4.

Also, you can visually see the pixel output on the smooth surface from a laser-based mouse (image contrast seen) versus an LED-based optical mouse (no contrast seen).

LED Optical Mouse   Laser Mouse

FIGURE 10.11
Design Concept 5.

nature of laser light creates patterns of high contrast when its light is reflected from a surface. The pattern appearing on the sensor reveals the details on any surface, even glossy surfaces that would look completely uniform when exposed to the LED incoherent illumination.

While an optical mouse offers a resolution of 800 dots per inch (DPI), replacing the LED with a laser increases the DPI to approximately 2000. The increased resolution translates to greater precision. Rather than producing a continual light beam, laser mice observe a standby or sleep mode in which the light dims when not in use and returns to full power when moved.

Concept 6—Plug Point Charger (see Figure 10.12): Making the mouse wireless involved directing its power supply to a rechargeable battery system for more convenience. The concept of a plug point at the bottom shell of the mouse was introduced. The battery located at the bottom of the mouse can be easily charged by plugging in the mouse to any electrical wall outlet. Trickle charging will be used so as to avoid overcharging or damaging the battery.

Concept 7—Touch-Sensitive Scroller: Remote Mouse (see Figure 10.13): This design concept will feature a touch-sensitive scroll panel. A swipe

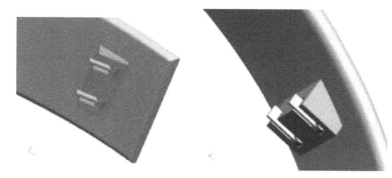

FIGURE 10.12
Design Concept 6.

FIGURE 10.13
Design Concept 7.

of the finger across the surface enables the inertial scrolling mechanism, which adjusts its speed according to the speed of the finger
swipe. The ratchet-scrolling mechanism retracts so the wheel can
spin with virtually no friction.

Free space motion control technology is designed to provide accurate, responsive navigation. This technology is based on a combination
of micro-electromechanical systems (MEMS) sensors, digital signal
processing (DSP) technology, and radio-frequency (RF) wireless technology that will allow a user to hold the mouse in any orientation,
point in any direction, and enjoy effortless, intuitive cursor control.

10.4.2.1 Pugh's Concept Selection

The seven different concepts were developed as described above, and Pugh's
Concept Selection matrix was used to compare the concepts and determine
the best fit. Figure 10.14 shows a simplistic illustration of Pugh's Concept
Selection matrix used to assist in selecting the best design.

From the developed concepts the team developed two hybrid design con-
cepts. Hybrid Concept 1 includes the features of Concepts 1, 2, 3, 4, 5, and 6.
Hybrid Concept 2 includes the features from Concepts 1, 2, 3, 5, 6, and 7.
Therefore, the selected concepts have the following features:

Hybrid Concept 1:
- Wireless
- Elimination of fasteners
- Modified clicker design
- Trackball concept
- Laser
- Plug point charger

Criteria	Concept 1—Wireless	Concept 2—Elimination of Fasteners	Concept 3—Modified Clicker Design	Concept 4—Trackball Concept	Concept 5—Laser	Concept 6—Plug Point Charger	Concept 7—Touch-Sensitive Scroller	Datum	Hybrid Concept 1 (1,2,3,4,5,6)	Hybrid Concept 2 (1,2,3,5,6,7)
Ergonomics	S	S	S	+	S	S	S	X	+	S
Price	+	+	+	+	+	+	+	X	+	+
Portability	S	S	S	S	S	S	S	X	S	S
Ease of charging	–	–	–	–	–	+	–	X	+	+
Aesthetics	–	S	S	–	S	S	S	X	S	S
Sensitivity of movement	S	S	S	S	S	S	S	X	S	S
Size	S	S	S	–	S	–	–	X	S	S
User friendliness	+	+	+	+	+	+	+	X	+	S
Sum of +	2	2	2	3	2	3	2	X	4	2
Sum of –	2	1	1	3	1	1	2	X	0	0
Sum of S	4	5	5	2	5	4	4	X	4	5

FIGURE 10.14
Pugh's Concept Selection matrix.

Hybrid Concept 2:

- Wireless
- Elimination of fasteners
- Modified clicker design
- Laser
- Plug point charger
- Touch-sensitive scroller

10.4.2.2 Detailed Model of the Product Design

The design concept of this mouse was selected by placing more emphasis on human ergonomics and human comfort, also keeping in mind the cost

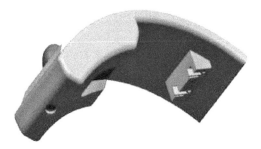

FIGURE 10.15
Hybrid final design.

and ease of assembly (see Figure 10.15). It is referred to as "THE ARC" for its shape completely resembles an arc.

Key advantages of this mouse include

1. The arc shape aids the customer in getting the proper grip. The arc shape is comfortable from an ergonomic point of view as the palm gets less fatigued with prolonged use.
2. It is lightweight making it the lowest of all the design concepts.
3. Much of the material has been removed as compared to the conventional mouse. Hence, less cost is incurred in manufacturing model.
4. The area of the mouse that moves over the pad or flat surface laser light area is quite less. Hence, the mouse can be easily moved due to lower contact area between the mouse and the pad.

10.4.3 Design Phase Case Discussion

1. How did you generate your design concepts?
2. How did you determine how your concepts compared using the Pugh Concept Selection matrix?
3. How did you derive the best combination of your design elements from each concept?

10.5 Optimize Phase

10.5.1 Optimize Phase Activities

1. Implement pilot process.
2. Assess process capabilities.
3. Optimize design.

10.5.2 Optimize

This phase optimizes the robustness of the technology under noise factors demanding more investment. Here, the design concept is finalized and reviewed. In addition, the influences of the factors on one another are compared and possible flaws are identified and analyzed.

Next, the noise factor diagram is created to understand the flow of noise. Noise factors such as external noise factors, deterioration noise factors, and unit-to-unit noise factors must be determined, and their effect is traced in the form of the output noise factor. Noise factor experiments have to be conducted to study the magnitude and directional effect on the output. A FMEA can be used to identify and screen noise factors. Analysis of variance (ANOVA) can be used to determine which noise factors are statistically significant. The results of the noise factor experiments can be used to understand how the noise factor effects can be optimized. Control factors that interact with the noise factors are good candidates to help reduce the sensitivity due to noise. Depending on the type of interaction (moderate, strong, weak, etc.) between control factors and noise factors, various control factors can act as robustsizing factors, adjustment factors, or economic factors.

Checklist: Design of experiments, failure mode effect analysis, analysis of mean, analysis of variance, design capability studies, critical parameter management

Scorecard requirements:

- Identifying noise factors and control factors from both customers and competitors
- Lowering occurrences of high-severity, high-occurrence defects
- Identifying factors that influence each other
- Determining probable solutions to failure

10.5.2.1 Design for X (DFX)

In the design for X methods, all of the design concepts were evaluated based upon the tools required for assembling the final product and the time taken in each concept to complete the assembling process (see Figures 10.16 through 10.19).

10.5.2.2 P-Diagram

The basic noise factors that interfere with the functioning of the device are described in the parameter chart. To avoid these noise factors during design, certain control factors are considered to improve the performance of the product as shown in Figure 10.20.

Item	Part Name	Method of Manufacture	Easy to Grasp	Easy to Align	Easy to Manipulate	Method of Fastening	Tools Required	Quantity
1	Left clicker	Injection molding	Yes	Yes	Yes	Mate with shield	No	1
2	Right clicker	Injection molding	Yes	Yes	Yes	Mate with shield	Yes	1
3	Mouse shield	Injection molding	Yes	Yes	Yes	Screwed to base	Yes	1
4	Roller	Injection molding	No	No	Yes	Mated with clickers	No	1
5	Screws	Standard part	No	No	No	Threaded	Yes	2
6	Circuit plate	Standard part	Yes	No	Yes	Mated with base	No	1
7	Mouse base	Injection molding	Yes	Yes	Yes	Screwed to shield	Yes	1
8	Optical light emitter	Standard part	Yes	No	Yes	Mated and sites on circuit plate	No	1
9	Receiver	Standard part	Yes	No	Yes	Mated with light emitter	Yes	1

FIGURE 10.16
DFX principles—chart for assembly.

Number	Part Name	Quantity	Part Design	Time/Part	Total Time
1	Left clicker	1	Bad	4	4
2	Right clicker	1	Bad	4	4
3	Mouse shield	1	Bad	4	4
4	Roller	1	Good	11.5	11.5
5	Screws	2	Bad	11.5	23
6	Circuits and plate	1	Good	8	8
7	Mouse base	1	Bad	4	4
8	Light emitter	1	Good	8	8
9	Receiver	1	Good	8	8

FIGURE 10.17
DFX principles—original assembly time.

10.5.3 Optimize Phase Case Discussion

1. How did you define a failure?
2. How did you determine which factors were significant to your design?
3. How did you determine the design solutions to prevent defects?

10.6 Validate Phase

10.6.1 Validate Phase Activities

1. Validate process.
2. Assess performance, failure modes, and risks.
3. Iterate design and finalize.

10.6.2 Validate Phase

In this phase, the team verified the final design parameters with responses from customers by providing them with prototypes.

Checklist: Measurement system analysis, manufacturing process capability study, reliability assessment, worst-case analysis, analytical tolerance design

Item	Part Name	Method of manufacture	Easy to Grasp	Easy to Align	Easy to manipulate	Method of fastening	Tools Required	Quantity
1	Clicker	Injection molding	Yes	Yes	Yes	Sit on groove on top cover	No	1
2	Top cover	Injection molding	Yes	Yes	Yes	Snap fit with base	Yes	1
3	Trackball Scroller	Standard Part	Yes	Yes	Yes	Sit on a circular sleeve inside the mouse	Np	1
4	Circuit Plate	Standard part	Yes	Yes	Yes	Mated with base	No	1
5	Mouse Base	Injection molding	Yes	Yes	Yes	Snap fit with top cover	Yes	1
6	Laser Light Emitter	Standard Part	Yes	No	Yes	Mated and sits on circuit plate	No	1
7	Receiver	Standard Part	Yes	No	Yes	Mated with light emitter	Yes	1
8	USB Dongle (Transmitter)	Standard Part	Yes	Yes	Yes	Mates with the bottom shell	No	1
9	Charger plug	Standard part	Yes	No	Yes	Molded to the bottom shell	Yes	1

FIGURE 10.18
DFX principles—chart for redesign assembly

Number	Part Name	Quantity	Part Design	Time/Part	Total Time
1	Clicker	1	Good	4	4
2	Top cover	1	Good	4	4
3	Trackball scroller	1	Good	4	4
4	Circuit plate	1	Good	4	4
5	Mouse base	1	Good	4	4
6	Laser light emitter	1	Good	8	8
7	Receiver	1	Good	8	8
8	USB dongle	1	Good	4	4
9	Charger plug	1	Good	8	8
				Note: Total Assembly Time = 48 seconds.	

FIGURE 10.19
DFX principles—redesign assembly time

10.6.2.1 Design Prototypes

A prototype design was developed of the hybrid concept as shown in Figure 10.21. These prototype designs were also shown to potential customers.

10.6.2.2 Failure Mode Effect Analysis

Brainstorming and a detailed study of the product helped in analyzing the prominent factors that cause failure to the product. The possible failures,

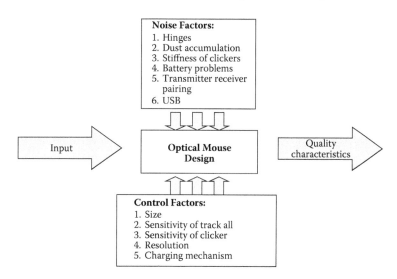

FIGURE 10.20
P-Diagram.

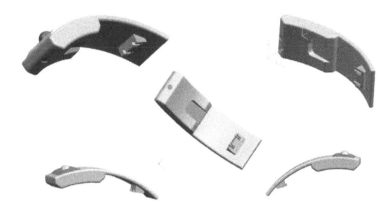

FIGURE 10.21
Design prototypes.

consequences, and their severity were determined by the failure mode effect analysis (FMEA) diagram as shown in Figure 10.22. Based upon this analysis the risk priority number (RPN) for each possible failure is calculated. From the usage of FMEA, considering the RPN factor, the team prioritized the possible potential failures to design a more efficient product.

From the FMEA analysis the design failures were prevented and solutions were found for the unavoidable failure modes.

In the verify phase, a prototype was created and was kept for customer evaluation. When compared with the basic model, customer satisfaction increased to 55% and the team was highly satisfied with the features of the product design. The statistics are given in Figure 10.23.

10.6.3 Validate Phase Case Discussion

1. How did you identify potential failure modes of your product?
2. How did you identify potential risks of your product?
3. How would you assess the potential market for your product?

10.7 Summary

Design for Six Sigma principles were used for the successful completion of the project. The basic project requirements were described through the use of VOC and VOT. Initially, several concepts were designed, and later all of the concepts were combined to finalize two hybrid concepts. The House of Quality was developed to ensure that the team's activity is in accordance with the voice of the customer.

							Page 1 of 1	
Item Design of optical mouse—Six Sigma project				Process Responsibility Soundar			Prepared by Soundar, Vishwa, Swathi, Priya	
Model Year(s)/Vehicle				Key Date 4/26/2010			FMEA Date (Orig) 04/20/2010	
Core Team: Soundar, Vishwa, Priya, Swathi								

Process Function Requirements	Potential Failure Mode	Potential Effect(s) of Failure	SEV	Potential cause(s)/ Mechanism(s) of Failure	Occ	Current Process Controls	Detec.	RPN	Recommended Action(s)
Parts	Scroller	Reduction in sensitivity of movement, Creeky sound	2	Dust accumulation, abrasion	2	None	4	16	Easy removal because of ease in disassemblement
	Hinges	Assembling issues	4	Wear and tear or rough usage	2	Hard material provided to prevent breakage	5	40	Careful reassembling
	Battery short circuit	Nonfunctioning mouse	5	Variable power supply, moisture and dust in atmosphere	3	Embodied in the design	3	45	
	USB	Mouse stops working	5	Power issues/not enough power provided to transmitter	2	None	2	20	Plug into powered USB port
	USB	Mouse stops working	5	Battery failure	2	None	2	20	Replace the batteries
	Clicker	No/slow response for clicking	4	Dust accumulation, misalignment	2	None	3	24	Dusting using compressed air

FIGURE 10.22
Design failure mode and effects analysis.

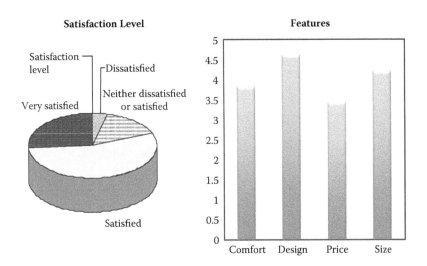

FIGURE 10.23
Prototype survey.

As a design team, implementing the DFSS principles to the product design enabled a savings of 26.5 seconds of assembly time per part, which is approximately 36% time saved per part. The design and the comfort levels of the product were increased along with its functional performance.

Thus, through the application of the DFSS methodology, a hybrid optical mouse was designed with minimal production failures and elimination of functional disorders. The product was designed per the customer requirements and was commercially viable with a nominal price. At the end of the process, the team was able to meet the project goals and the product was in line with the customer requirements from the surveys conducted.

11

Design of a Duffel Bag—A Design for Six Sigma Case Study

Jason Castro, Colby Krug, Patrick McCarthy,
Elizabeth Cudney, and Sandra Furterer

CONTENTS

11.1 Project Overview

Current duffel bags on the market contain the odor and moisture of the clothes stored within the bag so that it cannot escape the duffel bag and contaminates the surrounding area. Moisture leaks through the bag onto seats and other absorbent materials. Odor permeates the duffel bag and dissipates into the surrounding area such as your car or locker.

The Design for Six Sigma (DFSS) team will examine and recommend changes to the major process steps for the use of a duffel bag to decrease the odor and moisture of fitness gear stored in it. The investigation is limited to the moisture and odor absorption functions of the duffel bags.

The goal for the Design for Six Sigma team is to develop a duffel bag for people to store their wet and dirty clothes in, after they conduct physical fitness activities, that absorbs and dissipates the odor and moisture from the clothes. With all these features the bag should be cost effective using available technology. The proposed cost range is $20 to $40. The goal is to reduce moisture and odor of fitness clothes by 75%.

The requirements and expectations are to design the duffel bag based on voice of the customer (VOC) and use the DFSS methodology. The use of DFSS will result in a more efficient and organized design process.

11.2 Identify Phase

11.2.1 Identify Phase Activities

It is recommended that students work in project teams of three to four students throughout the DFSS case study.

1. *Develop Project Charter*: Use the information provided in the Project Overview section and the project charter format to develop a project charter for the DFSS project.

2. *Team Ground Rules and Roles*: Develop the project team's ground rules and team members' roles.

3. *Develop Project Plan*: Develop your team's project plan for the DFSS project.

11.2.2 Identify

11.2.2.1 Project Charter

The first step is to develop a project charter.

Project Name: Odor- and moisture-free duffel

Project Overview: The goal for the Design for Six Sigma team is to develop a duffel bag for people to store their wet and dirty clothes in after they conduct physical fitness activities that absorbs and dissipates the odor and moisture from the clothes.

Problem Statement: Current duffel bags on the market contain the odor and moisture of the clothes stored within so that it cannot escape the duffel bag and contaminates the surrounding area. Moisture leaks through the bag onto seats and other absorbent materials. Odor permeates the duffel bag and dissipates into the surrounding area such as your car or locker.

Customer/Stakeholders: Fitness personnel, military personnel, manual labor professionals. What is important to these customers is the need to store fitness apparel that does not release moisture or odor to the surrounding area (i.e., car, locker, office).

Goal of the Project: Reduce moisture and odor of fitness clothes by 75%.

Scope Statement: The Design for Six Sigma team will examine and recommend changes to the major process steps for the use of a duffel bag to decrease the odor and moisture of fitness gear stored in it. The investigation is limited to the moisture and odor absorption functions of the duffel bags.

Projected Financial Benefit(s): Based on an operating budget of $50, this project will attempt to avoid costs connected with cleaner costs in the region and disposal of moldy clothes.

11.2.2.2 Team Ground Rules and Roles

The team informally developed several ground rules for the project:

- Everyone is responsible for the success of the project.
- Listen to everyone's ideas.

FIGURE 11.1
Project Gantt chart.

- Treat everyone with respect.
- Contribute fully and actively participate.
- Be on time and prepared for meetings.
- Make decisions by consensus.
- Keep an open mind and appreciate other points of view.
- Communicate openly.
- Share your knowledge, experience, and time.
- Identify a backup resource to complete tasks when not available.

11.2.2.3 Project Plan

A Gantt chart was used to keep the project development cycle on track and ensure that all key steps in the process were accomplished as shown in Figure 11.1. This chart includes dates and tasks that are essential for appropriate usage of time management during the scheduled period of work.

11.2.3 Identify Phase Case Discussion

1. DFSS Project Charter: Review the project charter presented.

   a. A problem statement should include a view of what is going on in the business and when it is occurring. The project statement should provide data to quantify the problem. Does the problem statement provide a clear picture of the business problem? Rewrite the problem statement to improve it.

   b. The goal statement should describe the project team's objective and be quantifiable, if possible. Rewrite the goal statement to improve it.

   c. Did your project charter's scope differ from the example provided? How did you assess what was a reasonable scope for your project?

2. Project Plan

    a. Discuss how your team developed their project plan and how they assigned resources to the tasks. How did the team determine estimated durations for the work activities?

11.3 Define Phase

11.3.1 Define Phase Activities

1. *Collect VOC*: Create a VOC survey to understand the current and potential customers' requirements.

2. *Identify critical to satisfaction (CTS) measures and targets*: Based on the VOC, determine the CTS measures and then develop targets using benchmarking data.

3. *Translate VOC into technical requirements*: Using the CTS measures and targets, identify the technical requirements for the product.

11.3.2 Define

In this phase, the aim was to research the duffel bag market and gather customer requirements to capture the voice of the customer (VOC) through the use of an online survey. The survey was sent out to over one hundred people, and 69 responses were received ranging from every type of person who might use a duffel bag. These needs were then translated to useful metrics, prioritized, and evaluated to obtain a clear understanding of customer requirements and to help in the initial design of the bag.

    Checklist: Market segment analysis by visiting local sports stores and performing a market trend forecast to address the question, "Is there something like this out on the market yet?" In addition, benchmarking, gathering voice of the customer through survey, KJ analysis, quality function deployment (QFD), and Kano diagrams were utilized to ensure the team was heading in the right direction.

Scorecard requirements:

- Studying the market forecast
- Benchmarking our product to others available in the market
- Gathering information of voice of the customer through an online survey
- Translating customer needs to useful metrics and ranking them through building our House of Quality

- Prioritizing customer requirements based on survey results
- Performing Kano analysis
- Completing House of Quality

11.3.2.1 Identifying "Voice of the Customer"

In order to identify the needs of the customer, the team created a survey designed to help us understand the most important factors to the NODOR bag. The team developed this survey with the customer in mind, from those who would use this device regularly to those who would be able to purchase but did not have any experience in purchasing a product similar to ours in the past. The majority of customers we targeted were those who have some previous duffel bag purchasing experience. The greatest needs of the customer have been identified as

- 62% people preferred a medium-sized bag.
- 42% people preferred a bag that weighs less than 2 lbs (before clothes).
- 70% people would pay between $30 and $35 for this bag.
- 67% people would store the bag in their car.
- 67% people would use the bag at the gym.
- 67% people prefer a shoulder-carried bag.

The results of our survey (see Figure 11.2) led the team to develop different models based on a customer who prefers a medium bag, which he or she would like to use three to five times weekly at the gym and store in a vehicle, and for these requirements pay between $30 and $35.

11.3.2.2 Structuring and Ranking Customer Needs

After identifying all the wants and needs of the customer, we ranked and sorted them to realize the most important and critical needs as shown in Figure 11.3.

11.3.2.3 Analysis of Competitors in the Market

The next step was to identify our competition. Our product is semi-innovative; therefore, the process of identification was twofold. First, the team searched out duffel competition and then moisture wicking and defeating technologies. The team visited several sporting goods stores and analyzed different types of duffel bags ranging in price, size, and quality. Next, the

1. What size duffelbag do you use? Create Chart Download

	Response Percent	Response Count
a) Small - 11x20"	21.7%	15
b) Medium - 12x24" *Medium*	62.3%	43
c) Large- 14x35"	15.9%	11
answered question		69
skipped question		0

2. How heavy could the duffel bag be? (without clothes) Create Chart Download

	Response Percent	Response Count
a) Less than 2lbs *Less than 2*	42.0%	29
b) 3-4 lbs	34.8%	24
c) 5-6 lbs	18.8%	13
d) over 7lbs	5.8%	4
answered question		69
skipped question		0

3. How often do you work out a week? Create Chart Download

	Response Percent	Response Count
a) 1-2	29.0%	20
b) 3-4 *3-4 Times a Week*	39.1%	27
c) 5-6	20.0%	20
d) Everyday	5.8%	4
answered question		69
skipped question		0

4. How much would you be willing to pay for an odorless duffelbag? Create Chart Download

	Response Percent	Response Count
a) $30	34.8%	24
b) $35 *$35 Dollars*	36.2%	25
c) $40	21.7%	15
d) $45	10.1%	7
answered question		69
skipped question		0

FIGURE 11.2
Customer survey results.

team compared materials such as dry-fit, mesh, baking soda bags, and other technologies that may compete against it. There were only a few competitors the team believed would be competing for market space after our innovative entry.

*No direct comparison was available due to usage of varying technologies.

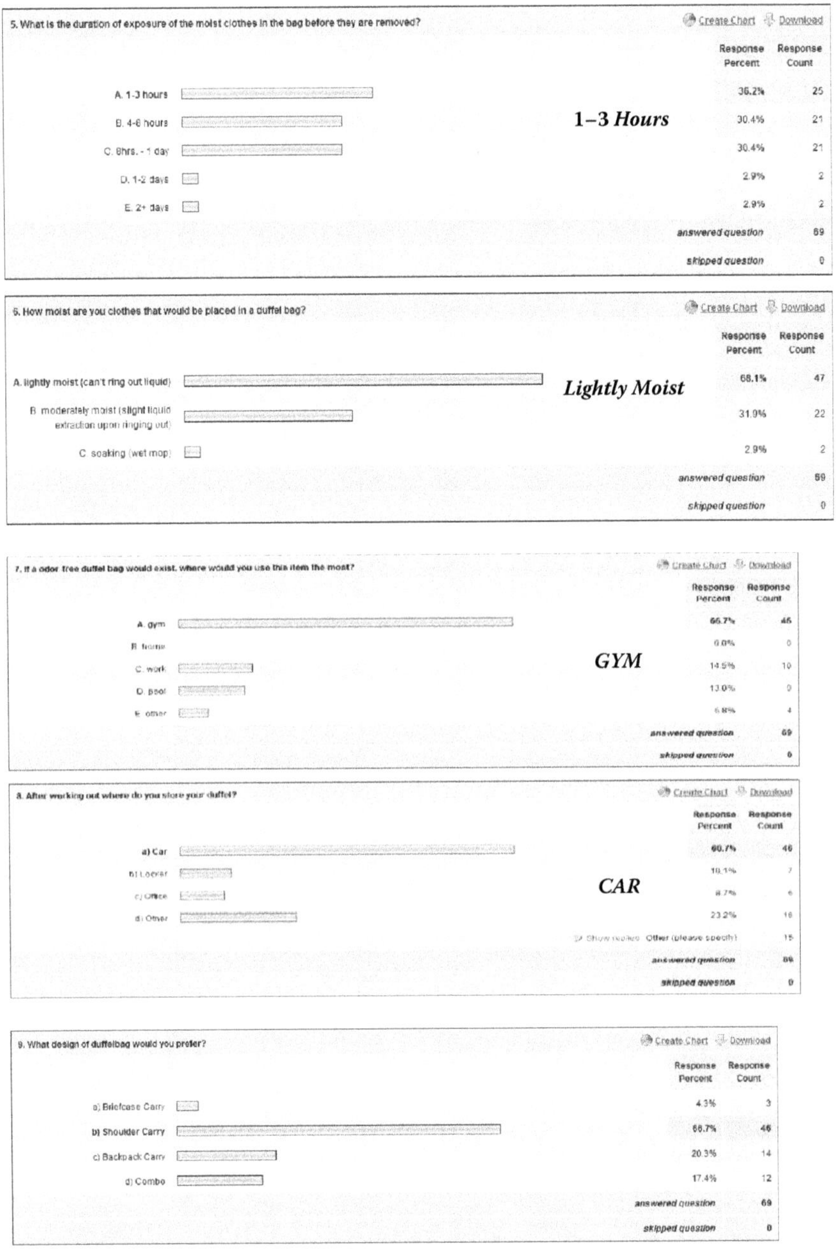

FIGURE 11.2
(Continued)

Number		Need	Importance
1	NODOR bag	What size duffel would you prefer?	5
2	NODOR bag	What is the ideal weight of the bag?	4
3	NODOR bag	How often do you work out a week?	2
4	NODOR bag	How much would you pay for a bag with said capabilities?	5
5	NODOR bag	What is the moisture level of the clothes you would use?	4
6	NODOR bag	Where would you use the bag?	3
7	NODOR bag	After working out, where would you store the bag?	3
8	NODOR bag	What is your preferred carrying method?	4
9	NODOR bag	Where would you store the bag?	3

FIGURE 11.3
Ranked customer needs.

11.3.2.4 Quality Function Deployment

What we wanted to do—Transform the VOC into engineering characteristics by prioritizing each product characteristic and all the while setting development goals for the entire product.

How we did it—(rank relationships) (i.e., each "must have" analyzed to ensure accomplished, then backwards plan to achieve manufacturing success)

What we used—House of Quality

Using the voice of the customer we built the House of Quality (see Figure 11.4). These are the steps we used to build it:

Step 1: List the customer requirements

Step 2: List the technical descriptors (characteristics that will affect more than one of the customer requirements, in development or production).

Step 3: Compare the two (customer requirements to technical descriptors) and determine relationships.

Step 4: Develop the positive and negative interrelated attributes and identify "trade-offs."

11.3.2.5 Product Design Metrics

The next step was developing the characteristics into metrics in order to identify our ideal specifications to meet customer requirements as shown in Figure 11.5.

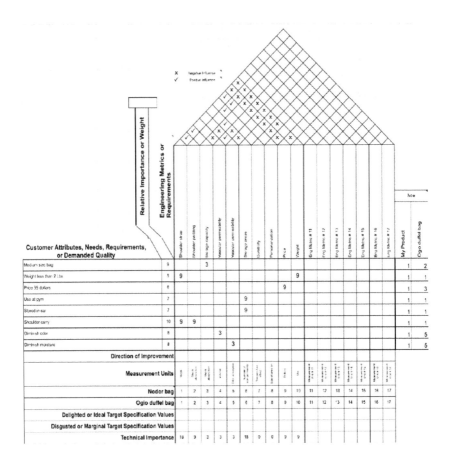

FIGURE 11.4
House of Quality.

11.3.2.6 Kano Analysis

For the design requirements, the team used the Kano analysis as shown in Figure 11.6. First, the team aligned the duffel bag to meet one-dimensional needs and their expected quality. Next, the team took the other features and developed these as "extra" or exciting quality features so it was easy to distinguish between the bag and the new technologies we were implementing. For the overall design of the NODOR bag we went back through the Kano analysis after the final design to ensure quality design.

The first process was to ensure the product met customer needs with the basic design. Those features are identified below:

1. A medium-size bag (12″ × 24″)
2. Shoulder carried
3. Weight of less than 2 lbs (empty)

Metric Number	Engineering Metrics	Measurement Unit	NODOR Bag	Standard Duffel Bags
1	Shoulder strap	Width	8	1
2	Shoulder padding	Shock absorption	7	2
3	Storage capacity	Volume	4	3
4	Vascular permeability	Odor absorption	2	4
5	Vascular permeability	Moisture absorption	1	5
6	Storage areas	Number of compartments	6	6
7	Durability	Temperature effect	5	7
8	Personalization	Size of area (in)	3	8
9	Price	Dollars	9	9
10	Weight	Lbs	10	10

FIGURE 11.5
Product design metrics.

The next step was to add exciting new qualities:

1. Mesh moisture-wicking liner
2. Fan
3. Scent disruption technology
4. Condensation removal powder (integrated in liner)

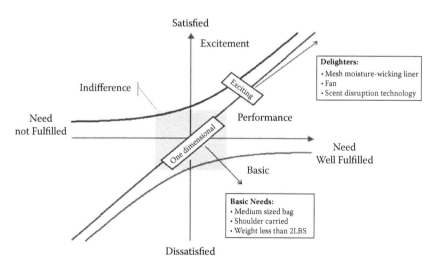

FIGURE 11.6
Kano analysis.

Again, the collection of data through the survey fed the ability to split the attributes between expected quality and exciting quality. The immediate reward for the customers is their expected quality. The key to the marketing strategy is the "extra" technology. This is the exciting quality that will sustain the product's niche in the market.

11.3.3 Define Phase Case Discussion

1. How did your team perform the VOC collection? How could VOC collection be improved?
2. Did your team create and distribute a customer survey, and if so, what is the appropriate statistical analysis to perform to identify the importance of the customers' requirements?
3. Did you perform a quality function deployment (QFD)? How did you identify the technical requirements and the correlations between customer and technical requirements?
4. What is the value of using the Kano model in your VOC analysis?

11.4 Design Phase

11.4.1 Design Phase Activities

1. Identify process elements.
2. Design process.
3. Identify potential risks and inefficiencies.

11.4.2 Design

In this phase, the team developed seven NODOR bag design concepts with regard to the customer requirements and evaluated each of those concepts to identify the best and most practically deliverable design that satisfies the customer's needs and our cost factors.

Checklist: Concept generation technique (The Theory of Inventive Problem Solving [TRIZ], brainstorming), design for manufacture and assembly, concept generation, affinity diagrams, Pugh concept evaluation and selection

Scorecard requirements:

- Generating seven design concepts that meet customer requirements
- Evaluating superior concepts and superior technology to beat the market

FIGURE 11.7
Concept 1.

- Analyzing and studying feasibility of superior concepts—looking at the competitors
- Adding value to the design by thinking "outside the box"

Bag 1: The prototype with just the tube sock baking soda absorber (see Figure 11.7)

Bag 2: A shoulder-carried duffel bag with a Fresh-pack liner/membrane that can be described as a bag with this liner boxing in the damp clothing (see Figure 11.8)

Bag 3: A shoulder-carried duffel bag with no membrane; however, it has an exhaust fan with a dryer sheet filter that can be described as the fan you would see underneath a laptop to keep it cool with an on/off switch (see Figure 11.9)

Bag 4: A shoulder-carried duffel bag with a baking soda membrane insert that sits on the floor of the duffel bag (see Figure 11.10)

Bag 5: A shoulder-carried duffel bag that is plain mesh with a disposable dryer sheet filter that encompasses the inside of the bag (see Figure 11.11)

Bag 6: A shoulder-carried duffel bag that has a fan on the side of it instead of on top (see Figure 11.12).

FIGURE 11.8
Concept 2.

FIGURE 11.9
Concept 3.

FIGURE 11.10
Concept 4.

FIGURE 11.11
Concept 5.

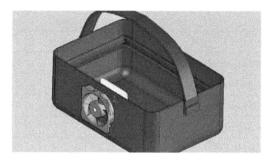

FIGURE 11.12
Concept 6.

> *Final Product*: A shoulder-carried duffel bag that has a well-padded shoulder strap. The bag will have a corrugated fresh-pack liner that will run along the floor of the bag with a disposable baking soda membrane encompassing the inside of the bag. It will have a lightweight exhaust fan that will have a dryer sheet filter and operate on battery power with an automatic shutoff of 10 minutes.

11.4.2.1 Pugh's Concept Selection

Through our product development the team created seven different designs of the NODOR bag. In order to determine the concept that had the greatest potential for success the team utilized the Pugh concept selection matrix to assist in narrowing the scope to a single design as shown in Figure 11.13. The team chose Concept 7 because it is currently unlike anything available in today's market and it provides the best results.

The selected design includes the following features:

- The bag dimensions will be 12″ × 24″.
- It will have a lining for freshness.
- The material of the bag will be a breathable fabric.
- The shoulder strap will be dual adjustable with a no-twist shoulder strap.
- It will have a baking soda bag insert.
- A fan cooling system will be installed with battery and plug-in capabilities.

11.4.2.2 Concept/Design Development

The design can be broken down into different modules (subsystems) according to the functions (no odor through air freshening, fan system, filter system, baking soda membrane) they perform.

	Concept						
Criteria	1	2	3	4	5	6	7
Medium-size bag	+	+	+	+	+	+	+
Weight less than 2 lbs	–	–	+	+	+	+	+
Price $35	–	–	–	–	–	–	+
Use at gym	+	+	+	+	+	+	+
Stored in car	+	+	+	+	S	+	+
Shoulder carry	S	S	S	S	S	+	+
Diminish odor	S	S	S	S	S	S	+
Diminish moisture	S	S	S	S	S	S	+
Sum of (+)	3	3	4	4	3	5	8
Sum of (–)	2	2	1	1	1	1	0
Sum of (s)	3	3	3	3	4	2	0

FIGURE 11.13
Pugh's Concept Selection matrix.

11.4.2.3 Detailed Model of the Product Design

Once the components were identified, the team developed a detailed model for the required design specifications.

11.4.3 Design Phase Case Discussion

1. How did you generate your design concepts?
2. How did you determine how your concepts compared using the Pugh Concept Selection matrix?
3. How did you derive the best combination of your design elements from each concept?

11.5 Optimize Phase

11.5.1 Optimize Phase Activities

1. Implement pilot process.
2. Assess process capabilities.
3. Optimize design.

11.5.2 Optimize

In this phase, the team reviewed the selected superior concept to identify possible flaws and their effects and probable solutions on how to reduce them. The team also compared the influence of one factor on another and outside noise factors.

Checklist: Design of experiments, failure mode effect analysis, analysis of mean, analysis of variance, design capability studies, critical parameter management

Scorecard requirements:

- Identifying noise factors and control factors from both customers and competitors
- Lowering occurrences of high-severity, high-occurrence defects
- Identifying factors that have influence with one another
- Determining probable solutions to failure

11.5.2.1 Design for X (DFX)

Design for X is a general term where X can mean any quality or cost criteria that affect the product. In the project the following DFX tools were used

- Design for Manufacturing (DFM)
- Design for Assembly (DFA)

11.5.2.2 Design for Manufacturing (DFM) and Design for Assembly (DFA)

After developing the final concept design the team decided the bag and the components will not be produced in house. However, the final assembly will be conducted in house. The components will be outsourced, and the product will be assembled in house.

Labor cost:
Assumed labor cost = $10 per hour

After the design for assembly (DFA) matrix was developed, the team finalized the sequence of operations to be performed to assemble the bag: *Does not take into context the bag itself, assuming an average bag is produced in 3 mins.

1. Time to attach fan to outer compartment—1 min 45 sec
2. Time to attach one filter to outside edges—25 sec (×2) = 50 sec
3. Time to attach mesh membrane—30 sec

Total material + Labor cost = $26.00 + $0.75 = $26.75

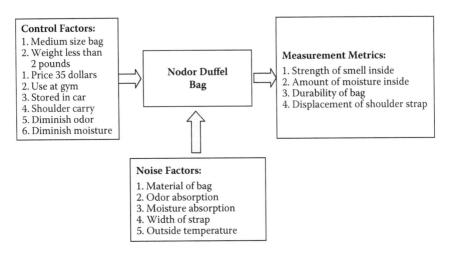

FIGURE 11.14
P-Diagram.

11.5.2.3 Design of Experiments

The team used design of experiments for the NODOR bag to understand the outputs that were affected by different inputs into the system. This method allowed the team to optimize results in targeted areas based on the results from the VOC and the required functioning parameters. The team determined the most cost-effective way to conduct product development.

The team used a parameter diagram (P-diagram) to identify areas that could be controlled based on the results from the voice of the customers and how they are affected by noise factors and combine to be analyzed by our measurement metrics (see Figure 11.14).

The effectiveness of the shoulder strap was easy to design based on simple weight distribution equations. The team was able to optimize the development of a strap that is a dual-adjustable, no-twist shoulder strap that uses a load distribution technology for optimal weight distribution, and its ergonomic design ensures comfort.

The team approached durability in the same manner and was able to research different materials that are available on the market. Cost analysis calculations were then conducted to find the best material that was affordable and still keep the price at $35. The team settled on using a waterproof, breathable, and durable fabric that is 10 times more durable than cotton duck, 3 times more durable than standard polyester, and 2 times more durable than standard nylon.

In developing the subsystems to diminish odor and moisture of clothes in the bag, the team took a current model and retrofitted it with two other methods to further diminish odor and moisture. We implemented a lining for freshness. The antimicrobial lining keeps the bag odor free. The team tested the bag's effectiveness through physical testing over a 6-hour period where the goal was to decrease the moisture in the bag by 75%. The lining only decreased odor by 50% and moisture by 10%. Therefore, the team placed a fan in the top compartment of the bag to increase circulation. The fan decreased moisture and odor by 60%. A baking soda sack was then incorporated that absorbed the remaining 15% of odor and moisture in the bag. The team was able to determine the size of the baking soda sack through calculating the amount of moisture that an ounce of baking soda could absorb.

The team also looked at the robustness of the project to determine if its short-term capability exists in the presence of noise factors and if it is tunable, meaning that the long-term capability for the bag exists in the presence of noise factors. In looking at the external noise factors of the temperature, the team cannot control this. It forced the team to treat the bag as a closed system and therefore not a significant factor in decreasing the odor and moisture in the bag. In looking at the internal noise factors such as absorption and material, there is more control and their effects are very tunable on the bag's required results. As stated above in the calculations, the team was able to determine the optimum material and diminishing capabilities to make the bag truly robust.

11.5.3 Optimize Phase Case Discussion

1. How did you define the design of experiment design to use?
2. How did you determine which factors and levels were significant to your design?
3. How did you determine the appropriate number of replications for your experiment?

11.6 Validate Phase

11.6.1 Validate Phase Activities

1. Validate process.
2. Assess performance, failure modes, and risks.
3. Iterate design and finalize.

11.6.2 Validate Phase

In this phase, the team verified the final design parameters with responses from customers by providing them with three prototypes. The team also arrived with feedback from various departments within the NODOR company to identify if this change would affect their performance both in the laboratory and if start-up costs could be covered.

> *Checklist*: Measurement system analysis, manufacturing process capability study, reliability assessment, worst-case analysis, analytical tolerance design
>
> *Scorecard requirements*:

- Prototype approved by customer
- Meets large portion of customer needs
- Ease to manufacture and cost estimate
- Reliability performance

The verification of the product design on Concept X is a critical element of the process. By modeling sensitivity variations across the product line, a level of confidence may be gained from the concept design with regard to potential commercialization. Specifically, for the NODOR bag, modeling variance may be conducted by using simple linear regression analysis and analysis of variance (ANOVA). The metric associated with ANOVA remains the S/N ratio.

Building on the optimization model developed earlier, the objective maximization of the S/N ratio and establishing a coefficient of determination R^2 that checks testing data to a functional relationship. In this case, the team will use a scatterplot of the percent of moisture versus time.

11.6.2.1 Verification Setup

It is known that the fan and baking soda bag reduce moisture significantly; however, it is not known if the rate of decrease of the moisture level represents a linear function. Once the team has received data for the functional performance of the final concept, they will verify the time the odor and moisture diminish in time in a test setup. The setup is for the R^2 testing shown in Figure 11.15.

Next, when the variance of moisture percent decrease with regard to time is credibly established, an ANOVA experiment is set up based on an orthogonal array to determine how the NODOR bag reacts under control parameter settings. An L_4 orthogonal array is chosen to determine how noise influences the system. Figure 11.16 shows the array and control parameters that the team set up for the experiment. This is identified by the team as future work.

S/N is computed in the larger, better format and is shown in the equation below:

$$S/N_{LTB} = -10\log\left[\frac{1}{n}\sum_{i=1}^{n}\left(\frac{1}{y_i^2}\right)\right]$$

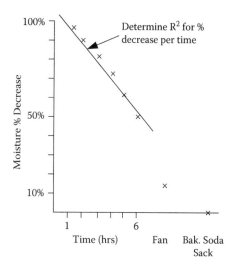

FIGURE 11.15
Verification setup.

ANOVA is then conducted by determining the grand total sum of squares (GTSS):

$$GTSS = \sum_{i=1}^{n}(S/N)_i^2$$

The GTSS equation will then lead to a decomposition of the sum of squares about the mean and will lead to the percentage contribution that the fan, baking soda, and timed exposure plays when exposed to hot or cold environments.

11.6.3 Validate Phase Case Discussion

1. How did you identify potential failure modes of your product?
2. How did you identify potential risks of your product?
3. How would you assess the potential market for your product?

11.7 Summary

Design for Six Sigma was applied to design a duffel bag used for storing workout clothes which reduced odor and moisture. Voice of customer requirements were collected to identify the customer requirements. QFD

Control Parameter	Level	
	1	**2**
Time	1 hour	6 hours
Fan	Off	On
Baking soda	Inactive	Active

Run	Time	Fan	Baking Soda	Temperature Noise		Mean	Standard Deviation	S/N
				Hot (80°F)	Cool (400°F)			
1	1	1	1	Moisture data	Moisture data			
2	1	2	2	Moisture data	Moisture data			
3	2	1	2	Moisture data	Moisture data			
4	2	2	1	Moisture data	Moisture data			

Mean
Standard Deviation

FIGURE 11.16
Design of experiments setup.

and Kano analysis were used during the Define phase to relate the customer requirements to the technical requirements. We defined the conceptual designs using the Pugh Concept Evaluation and Selection matrix. We prototyped our design for the selected design concept. We applied design of experiments to optimize our design, and validated our final design using the S/N ratio from our ANOVA analysis.

Section II

Service Process Design for Six Sigma Projects

12

Design of Women's Center Service Processes—A Design for Six Sigma Case Study

CONTENTS

12.1 Project Overview

An acute-care hospital recently embarked on building and designing a new Women's Center to respond to the exponential growth of women's health-care needs in their market area by redefining, refocusing, reconfiguring, and expanding its women's health services line. They had an existing smaller center with fewer services that was part of their outpatient health-care. It is recognized that a "one-stop shopping" model where women can receive a comprehensive range of services conveniently, comfortably, efficiently, and with speed and ease is desirable.

The goal was to provide women's services in a spa-like environment. The Women's Center will provide the comprehensive list of services shown in Figure 12.1. The services will be phased in as the facility is built and funds become available.

12.1.1 Project Goals

The intent of the Women's Center (WC) is to offer a diverse array of women's outpatient services in one location that provides convenient and easy access to multiple services including physician, diagnostic, ancillary, education, and outreach services. The vision of the WC is to build patient loyalty among women in the community and become the provider of choice for all their health-care needs. It is expected that the WC will not only increase market share for outpatient services but will also increase inpatient volume and referrals back to the hospital.

Primary care	Gynecology	Diagnostic imaging
Urogynecology	Oncology	Breast surgery
General surgery	Endocrinology	Rheumatology
Cardiology	Osteoporosis treatment	Bone density screenings
Medical spa	Meditation area	Lifestyle center
Education center	Boutique	Café

FIGURE 12.1
Women's Center Services.

The new Women's Center plan assumes *potential* process and staffing changes, such as

- Having radiology results available while the patient is in the center
- Providing the ability to assess authorization of additional diagnostics
- Being able to meet with a physician if the patient requests

These process requirements need to be clearly defined before the new process and staffing are defined. The requirements significantly impact the processes. This project focuses on designing efficient processes that meet the needs of the customers (women) who will receive services at the Women's Center. The processes should have work flows that optimize throughput, reduce wait times, and provide next-business-day results to the patients who receive women's services.

12.2 Identify Phase Activities

It is recommended that students work in project teams of three to four students throughout the Design for Six Sigma (DFSS) case study.

1. *Develop Project Charter*: Use the information provided in the Project Overview section to develop a project charter for the DFSS project.
2. *Perform Stakeholder Analysis*: Perform a stakeholder analysis, identifying project stakeholders.
3. *Develop Project Plan*: Develop your team's project plan for the DFSS project.

12.2.1 Identify Phase

12.2.1.1 Create Project Charter

The first step was to develop a project charter.

> *Project Name*: Design of Women's Center Service Processes
>
> *Project Overview*: Many outpatient facilities are focusing on providing comprehensive services to women in a comfortable setting. In a qualitative study of women who had received a mammogram in the prior 3 years, without a history of cancer, satisfaction was related to the entire experience, not just the actual mammogram procedure. The authors found seven satisfaction themes from the focus groups: (1) appointment scheduling, (2) facility, (3) general exam, (4) embarrassment, (5) exam discomfort/pain, (6) treatment by the technologist, and (7) reporting results (Engelman, Cizik, and Ellerbeck, 2005). This supports the focus of designing a seamless experience for women in the Women's Center through applying the Design for Six Sigma methodology and tool results.
>
> *Problem Statement*: On average, patients who visit the former Women's Center are transmitted to the primary care provider (PCP) within 57 hours. There was large variability in how and when the test results were communicated to the patients. Some of the questions that the team had to answer before the new center opened were as follows:
>
> - How do we measure the success of the new women's center?
> - What processes will stay the same for the new center?
> - What processes will be different in the new center?
> - What is our current state and future state?
> - Should we perform activities in parallel or series? Or both?
> - What is our forecasted demand? Do we have different processes based on our volume?
> - What is our goal for waiting time? Is it different for the different areas (i.e., registration, radiology reception, gowned waiting room)?
> - What should we prioritize patients by (i.e., appointment time, arrival time, waiting time)?
> - What is an acceptable wait time when having two tests done?
> - Will patients have their results available after their tests? Will the results be available for all types of tests (diagnostic/screening)?
> - Will patients have the ability to meet with a physician given the test result comes back requiring further testing?
> - Will patients be able to get a diagnostic test?

Customers/Stakeholders:

> *External:* Patients (women who receive services in the center), referring physicians, payers, donors

> *Internal:* Imaging technologists, radiologists, administration, registration, centralized scheduling, information technology, physicians, marketing and development

Goal of the Project: To design the processes and define the metrics that result in optimal flow for the new Women's Center Phase 1.

Scope Statement: The scope includes the new Women's Center diagnostic services (mammography, ultrasound, stereotactic biopsy, and bone densitometry), registration, and appointment scheduling.

Projected Financial Benefit(s): Patient satisfaction, increased capacity due to efficient workflow, and resultant revenue are potential financial benefits of this project.

12.2.1.2 Perform Stakeholder Analysis

The stakeholder analysis was performed to identify the project stakeholders. The stakeholder analysis definition is shown in Figure 12.2.

12.2.1.3 Create Project Plan

A Gantt chart was used to keep the project development cycle on track and ensure that all key steps in the process were accomplished. The Project Plan is shown in Figure 12.3. It shows the tasks, task durations, start and end dates, predecessors of each task, and the resources to perform each task. The project started in March 2010 and finished in November 2010.

12.2.2 Identify Phase Case Discussion

1. DFSS Project Charter: Review the project charter presented.

   a. A problem statement should include a view of what is going on in the business and when it is occurring. The project statement should provide data to quantify the problem. Does the problem statement provide a clear picture of the business problem? Rewrite the problem statement to improve it.

   b. The goal statement should describe the project team's objective and be quantifiable, if possible. Rewrite the goal statement to improve it.

   c. Did your project charter's scope differ from the example provided? How did you assess what was a reasonable scope for your project?

Stakeholders	Who Are They?	Potential Impacts/Concerns
Patient (women)	Receive services in the center	• Customer service • Quality • Efficiency
Referring physicians	Refer patients to the center and communicate results to the patient	• Quality of care
Payers	Pay for services, such as Managed Care, Medicare employers, self-pay	• Quality • Cost effective care across the continuum
Donors	People who donate money to the center	• Quality of care • Meet patient requirements
Imaging technologists	Work in the new Women's Center and perform procedures	• Patient satisfaction • Improved/well-designed work environment • Associate satisfaction
Radiologists	Read and provide results of imaging procedures	• Reduced volume and revenue • Physician satisfaction
Administration	Manage the hospital and center	• Volume • Revenue • Patient satisfaction • Physician satisfaction productivity
Patient access and centralized scheduling	Registration who registers patients and performs insurance authorizations, centralized scheduling who make the patient appointments	• Volume • Productivity • Timeliness of processes
Information technology	Provide phone and computer systems	• Meet requirements • On time • On budget
Physicians	Provide women's services	• Reduced volume and revenue • Physician satisfaction
Marketing and development	Perform business development, marketing, and fund raising	• New business • Funds available • Able to reach customers

FIGURE 12.2
Stakeholder analysis definition.

Task		Duration	Start Date	End Date	Predecessors	Resources
New Women's Center throughput Project	100%	174 days	3/17/10	11/12/10		
Create milestones	100%	1 day	3/17/10	3/17/10		
Identify Phase	100%	42 days	3/17/10	5/13/10		
Define project charter	100%	14 days	3/17/10	4/5/10		R., V.
Complete risk matrix	100%	6 days	3/24/10	3/31/10		All
Complete stakeholder analysis definition	100%	11 days	4/1/10	4/15/10	4	All
Develop communication plan	100%	12 days	4/28/10	5/13/10	4	R., A.
Develop project plan	100%	4 days	3/17/10	3/22/10	4	M.
Develop responsibilities matrix	100%	4 days	4/28/10	5/3/10	4	
Establish ground rules	100%	2 days	4/7/10	4/8/10		
Define Phase	100%	84 days	4/16/10	8/11/10	3	
Develop data collection plan	100%	3 days	6/30/10	7/2/10		M.
Gather voice of customer (VOC)	100%	20 days	6/14/10	7/29/10	12	All
Complete value stream map	100%	6 days	6/14/10	6/21/10		E.
Perform process mapping	100%	10 days	4/16/10	8/11/10	13	M., V.
Operational definitions	100%	2 days	6/30/10	7/1/10		V.
Establish baseline	100%	4 days	7/6/10	7/9/10	16	R.
Perform benchmarking	100%	10 days	7/6/10	7/19/10	12	J.
Design Phase	100%	62 days	6/9/10	9/2/10	11	
Strength–weakness-opportunity-threat (SWOT)	100%	7 days	6/9/10	6/17/10		A.
Quality function deployment (QFD)	100%	12 days	8/12/10	8/27/10		All
5 Why's	100%	16 days	8/12/10	9/2/10		V.
Cause/effect diagram	100%	7 days	8/12/10	8/20/10		M.
Optimize Phase	100%	77 days	6/10/10	9/24/10		
Quality function deployment	100%	12 days	9/3/10	9/20/10	19	All
Establish performance targets, project score	100%	20 days	6/10/10	7/7/10		V.
Optimize value stream	100%	15 days	8/16/10	9/3/10		All
Gain approval to implement (buy-in from stake)	100%	5 days	9/6/10	9/10/10	27	V.
Develop implementation plan	100%	5 days	8/16/10	8/20/10		M.
Develop communication plan	100%	6 days	8/16/10	8/23/10		A.
Develop training plan	100%	3 days	8/16/10	8/18/10		R.
Pilot	100%	30 days	8/16/10	9/24/10		All
Cost benefit analysis	100%	6 days	8/16/10	8/23/10		A.
Procedures	100%	11 days	8/16/10	8/30/10		V.
Validate	100%	49 days	9/8/10	11/12/10		
Validate processes	100%	26 days	9/27/10	10/30/10	24	All
Statistical analysis	100%	5 days	11/1/10	11/5/10	36	E.
Assess performance	100%	10 days	11/1/10	11/12/10	36	All
Dashboard, scorecards, hypothesis test	100%	21 days	9/8/10	10/6/10		All

FIGURE 12.3
Project plan.

2. Project Plan
   a. Discuss how your team would develop their project plan and how they assigned resources to the tasks. How would the team determine estimated durations for the work activities?

12.2.3 Define Phase Activities

1. *Collect voice of the customer (VOC)*: Create a VOC survey to understand the current and potential customers' requirements.
2. *Identify critical to satisfaction (CTS) measures and targets*: Based on the VOC, determine the CTS measures and then develop targets using benchmarking data.
3. *Translate VOC into technical requirements*: Develop a quality function deployment (QFD) House of Quality matrix relating the customer requirements to the potential technical or process requirements.

12.2.4 Define Phase

The following is a written report of the Define phase for the project, including the key deliverables developed as part of the prior exercises. The Define phase of the Identify-Define-Design-Optimize-Validate (IDDOV) process is designed to gain information on the voice of the customer (VOC) to understand the needs of the customers and begin translating those customer requirements into the processes' technical elements. The main activities of this phase are to collect VOC; identify CTS measures and targets; and translate VOC into technical requirements.

12.2.4.1 Collect Voice of the Customer

We first defined the market opportunities through a literature search and identified the following findings that support the importance of a women's center:

Market Opportunities:
Specific areas in which women have high mortality and morbidity rates include

- 60% of hospital patients are women
- 50% of all women over age 50 will have osteoporosis
- Hypertension is three times higher for women than men
- Depression is more common in women than men, affecting over 7 million women in the United States
- 75% of Alzheimer's patients are women

- 90% of rheumatoid arthritis cases are women
- 90% of patients with lupus are women
- An estimated 10 million women in the United States suffer from urinary incontinence
- Women have twice the incidence of multiple sclerosis than men
- 1 in 7 women age 45 or older has cardiovascular disease, and 45% of all female deaths are cardiac related
- Almost 80% of fibromyalgia cases are female
- Over 80% of bariatric surgery patients are women

The team then performed a literature search to identify customer requirements that impact satisfaction in a women's center. Engelman, Cizik, and Ellerbeck (2005) performed focus groups to identify the factors and dimensions related to a patient's experience with a mammography procedure. They identified the number of coded text lines from the focus group transcripts for the factors and dimensions related to the patient experience. We developed a Pareto chart identifying the most important factors and the most important dimensions for each factor. The Pareto chart for the factors is shown in Figure 12.4. The percentage importance based on the number of coded text lines for the dimensions of each factor are shown in Figures 12.5 through 12.11.

Scheduling was the highest importance factor, with the most important dimensions being scheduling convenience, reasons to schedule a mammogram, and financial issues around paying for a mammogram. Results was

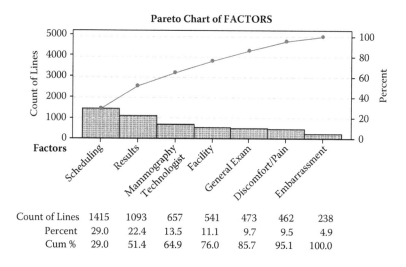

	Scheduling	Results	Mammography Technologist	Facility	General Exam	Discomfort/Pain	Embarrassment
Count of Lines	1415	1093	657	541	473	462	238
Percent	29.0	22.4	13.5	11.1	9.7	9.5	4.9
Cum %	29.0	51.4	64.9	76.0	85.7	95.1	100.0

FIGURE 12.4
Pareto chart of importance by factor.

Factors	Dimensions	Number of Lines	Definition	Importance
Scheduling	Convenience	529	Convenience of scheduling an appointment	37%
Scheduling	Reasons	306	Reasons to schedule or not to schedule a mammogram	22%
Scheduling	Financial	213	Financial issues surrounding paying for the mammogram	15%
Scheduling	Scheduling timing	187	Time between scheduling of appointment and actual appointment date	13%
Scheduling	Reminders	129	Reminders or other cues to schedule a mammogram	9%
Scheduling	Transportation	51	Transportation to/from the appointment	4%
		1415		

FIGURE 12.5
Results dimension.

the next most important factor, with the important dimensions being the manner in which the results are reported, the time it takes to receive the results, and the need to return for suspicious findings or additional procedures. The mammography technologist factor was the next most important element, with the important dimensions being the technologist's attitude toward the woman during the exam, the provision of the exam instruction, and the technologists' skill in performing the procedure. The facility was the next most important factor, with the most important dimension including the environment of the waiting room including cleanliness, having an educational video and magazines, noise level, décor, and beverage

Factors	Dimensions	Number of Lines	Definition	Importance
Results	Notification	358	Method in which results are conveyed	33%
Results	Results timing	249	The time it takes to receive notification of results and anxiety while waiting for results	23%
Results	Return procedures	221	Suspicious findings resulting in repeat views or returning for other procedures	20%
Results	Questions	156	Provision of information or education to address abnormal results	14%
Results	Relief	55	Feeling of relief when test is normal	5%
Results	Misconception	54	Misconception that technologist reporting adequate film quality was providing results of the exam	5%
		1093		

FIGURE 12.6
Results dimension.

Factors	Dimensions	Number of Lines	Definition	Importance
Mammography technologist	Attitude	215	Technologist attitude toward woman during procedure	33%
Mammography technologist	Exam instruction	183	Provision of exam instruction— necessary physical contact, breast placement	28%
Mammography technologist	Skill/experience	149	Technologist skill and experience in performing mammograms (good/bad)	23%
Mammography technologist	Breast positioning	110	Woman's perception of technologist's positioning of her breast (gentle/rough)	17%
		657		

FIGURE 12.7
Mammography technologist dimension.

availability. The general exam was the next most important factor, with the issues surrounding the gown worn for the procedure being the most important, followed by instructions to prepare for the procedure, and the cold or warm machine plate. Discomfort and pain was the next most important factor, with the most important dimension including uncomfortable pressure due to the breast plate compression. The least most frequently discussed element of mammograms is the embarrassment factor, including the most important dimensions of the embarrassment from participating in an exam of a personal nature; embarrassment of discussing the mammogram with family, friends, or their health-care provider; and self-consciousness due to being modest.

Factors	Dimensions	Number of Lines	Definition	Importance
Facility	Waiting room	281	Environment of waiting room at facility (cleanliness, educational video, magazines, noise, décor, beverages)	52%
Facility	Distance traveled	77	Distance traveled by woman to reach the facility	14%
Facility	Changing area	73	Place where woman disrobes in preparation for procedure and where personal effects are stored	13%
Facility	Exam room	68	Procedure/exam room environment	13%
Facility	Parking	42	Convenience of parking	8%
		541		

FIGURE 12.8
Facility dimension.

Factors	Dimensions	Number of Lines	Definition	Importance
General exam	Gown	170	Issues surrounding the gown worn during exam	36%
General exam	Preparation instructions	90	Instructions provided to prepare for mammogram (deodorant, jewelry, caffeine intake)	19%
General exam	Plate temperature	89	Cold or warm machine plate	19%
General exam	Safely/reliability	66	Questions about safety and reliability of procedure	14%
General exam	Unfamiliarity/ apprehension	30	Unfamiliarity with mammogram procedure leads to feelings of apprehension, fear, or being scared	6%
General exam	Forget experience	28	Unpleasant exam causes woman to block it out or want to hurry through the exam	6%
		473		

FIGURE 12.9
General exam dimension.

Factors	Dimensions	Number of Lines	Definition	Importance
Discomfort/pain	Pressure	118	Uncomfortable pressure due to breast plate compression	26%
Discomfort/pain	Skin stretch	83	Breast placement and plate pressure cause breast and surrounding skin to pinch, pull, stretch, or tug	18%
Discomfort/pain	Horror stories	74	Stories about bad mammogram experiences from friends and/or family members	16%
Discomfort/pain	Breast size	63	Size of breast determines level of pain experienced	14%
Discomfort/pain	Height of machine	56	Short stature causes uncomfortable extension of breast up to machine	12%
Discomfort/pain	Breast cysts	35	Fibrocystic disease leads to increased pain during mammogram	8%
Discomfort/pain	Anticipation/fear	33	Anticipation of pain results in fear or anxiety for mammogram procedure	7%
		462		

FIGURE 12.10
Discomfort/pain dimensions.

Factors	Dimensions	Number of Lines	Definition	Importance
Embarrassment	Related to culture	89	Embarrassment of participating in exams of a personal nature due to cultural background	37%
Embarrassment	Mammogram discussion	62	Embarrassment surrounding discussing mammograms with family, friends, or health-care providers (due to cultural background)	26%
Embarrassment	Modesty	54	Self-conscious during appointment due to being modest	23%
Embarrassment	Privacy	27	Exam environment not conducive to privacy	11%
Embarrassment	About body or breast size	6	Self-conscious about body image and/or large/small breast size	3%
		238		

FIGURE 12.11
Embarrassment dimension.

12.2.4.2 Identify Critical to Satisfaction (CTS) Measures and Targets

The Women's Center is fortunate to have some very special donors and supporters. Marketing held several focus groups to understand their critical requirements, so that we could extract the CTS criteria from the focus group qualitative data. The CTS are shown in Figure 12.12.

12.2.4.3 Translate Voice of the Customer (VOC) into Technical Requirements

12.2.4.3.1 Quality Function Deployment

Using VOC we built three Houses of Quality. These are the steps we used to build it:

Step 1: List the customer requirements with their importance rating on a scale of 1 (low) to 10 (high).

Step 2: List the technical process requirements that will affect more than one of the customer requirements, in the process.

Step 3: Compare the two (customer requirements to the technical customer requirements) and determine the relationships, identifying a strong relationship (9), medium relationship (3), low relationship (1), or no relationship (blank).

Step 4: Obtain the weighting and ranking for the process requirements.

Step 5: Carry the weighting to the second House of Quality. Place the process requirements on the left vertically, and identify the process components that will help to meet the process requirements.

Step 6: Determine the relationships between the process requirements and the process components.

Critical To Satisfaction
Environment
Comfortable
Ease of parking
Aesthetic rooms and geared psychologically to women
Creative access points to modern facilities
Guided through health issues in a nurturing, relaxing environment
Operational
Efficiently, with speed
Ease
Easy navigation throughout the system
Time-saving, convenience
Combined appointments
Same day results
Caring and competent professional staff
Customer focused amenities
Advanced technology as an enabler of superior care
One Visit, One Stop
Guided through health issues in a nurturing, relaxing environment
Service
Seamless integration of service components
All ages and life phases
Comprehensive range of services
One Visit, One Stop
Guided through health issues in a nurturing, relaxing environment
Coordination of medical and health concerns and treatments, multi-disciplinary team
Integrating between physicians, diagnostics, and ancillary
Functional medicine that combines traditional and integrative
Gender-specific medicine
Holistic care
Aligned with core values

FIGURE 12.12
Critical to satisfaction (CTS) criteria.

Step 7: Carry the weighting to the third House of Quality. Place the process components on the left vertically, and identify the process metrics that measure the process components.

Step 8: Determine the relationships between the process components and the process metrics.

House of Quality 1 is shown in Figure 12.13. House of Quality 2 is shown in Figure 12.14. House of Quality 3 is shown in Figure 12.15.

Quality Function Deployment—House of Quality 1

Process Requirements

Critical to Satisfaction (from Business Plan)	Importance	Valet Parking	Close in Parking	Interior Design	Service Levels of Process Times	Design for Six Sigma	Concierge	Patient Navigator	Warm Transfer between Central Scheduling and Doctor's office	VIP Doctor's office Service	Next-Day Results	Same-Day Appointments for Walk-Ins	Patients with no Physician, Referred to Doctor	Ability to Seamlessly Register Patients in information system	Referring Physician Preference Cards	Navigator Database	Provide Spiritual Care	Connect Patient to Cancer Center	Direct Patient to Financial Assistance, if Needed	Ability to Receive Films/Records	Provide Films/Records to Second Doctor	Provide Surgery	View VIP Status in information system	View VIP Status in information system for tracking	Same-Day Results for VIP and Girlfriends, and Request	Same-Day Screening, Diagnostic, Other Procedures for VIP, Other as Requested	Spiritual Care Hotline	Trained and Certified Staff	Women's Services	Superior Imaging Technology
Environment																														
Comfortable	7	9		9																								3		
Ease of parking	1		9	9						9																				
Aesthetic rooms and geared psychologically to women	10			9		1				3																				
Creative access points to modern facilities	7	1					3	9	1	1	1	3				1		1	1	1					1		1		3	1
Operational																														
Efficiently, with speed	10	1	1		9	9	1	1	3	3	1	3	3	3				1	1	1					9	9		3	9	3
Ease	9	9	3	3		3	9	9			1	3	1	9		1				3					9	9			3	
Easy navigation throughout the system	10	1					9	9		3					1	3														
Time-saving, convenience	8	9	3		3	9	9		3	3	3	3				9		3	1	3			1	1	9	9		3	3	9

Continued

Quality Function Deployment—House of Quality 1
Process Requirements

Critical to Satisfaction (from Business Plan)	Importance	Valet Parking	Close in Parking	Interior Design	Service Levels of Process Times	Design for Six Sigma	Concierge	Patient Navigator	Warm Transfer between Central Scheduling and Doctor's office	VIP Doctor's office Service	Next-Day Results	Same-Day Appointments for Walk-Ins	Patients with no Physician, Referred to Doctor	Ability to Seamlessly Register Patients in information system	Referring Physician Preference Cards	Navigator Database	Provide Spiritual Care	Connect Patient to Cancer Center	Direct Patient to Financial Assistance, if Needed	Ability to Receive Films/Records	Provide Films/Records to Second Doctor	Provide Surgery	View VIP Status in information system	View VIP Status in information system for tracking	Same-Day Results for VIP and Girlfriends, and Request	Same-Day Screening, Diagnostic, Other Procedures for VIP, Other as Requested	Spiritual Care Hotline	Trained and Certified Staff	Women's Services	Superior Imaging Technology
Combined appointments	8				3	9	3	3	3	3	3	9		1									1	1				1	3	
Same-day results	10					9	3	3	9	1	9		1			9	3	3	3				1	1	9			9	1	
Caring and competent	10						3	3		9																			9	
Customer-focused amenities	9	9		9			1			9	9	9					1	1	1	3			1	1	9	3	3		3	3
Advanced technology as an enabler of superior care	8					3	1	3						3															9	9
One visit, one stop Service	10	1			3	3	3	3	3	9		9	3	3			3	3	1	9					9	9		1	3	3
Seamless integration of service components	8				3	9		3	3	3	3	9	3	9			3	3	3	9					9	9		1		
All ages and life phases	6			3						9																				
Comprehensive range of services	8						3	3		3	1	1	1				1	1	1						9	9	1		9	

Customer requirement	Importance	C1	C2	C3	C4	C5	C6	C7	C8	C9	C10	C11	C12	C13	C14	C15	C16	C17	C18	C19	C20	C21	C22	C23	C24	C25	C26	C27	C28	C29
Guided through health issues in a nurturing, relaxing environment	9	3	3				9	9	3	3	3	3		1		9				1					3	3	1	3	3	3
Coordination of medical and health concerns and treatments, multidisciplinary team	8			3		3	3	3	3	3			1	3			1	1								9	1		9	9
Integrating between physicians, diagnostics, and ancillary	8			3		3	3	3	3	3			1	3			1	1								9	1		9	9
Functional medicine that combines traditional and integrative	7	1	1	3		9	9	9	9	9	9	9	1	1			1	1	1											
Gender-specific medicine	8			1	1		1	9		9													1	1						
Holistic care	9	3		3			3										1	1												
Aligned with core values	10							3	3	1							9	3	9								3		1	
Weighting		312	77	335	192	487	642	685	331	801	367	488	109	333	10	289	222	202	193	266	0	0	70	70	610	747	105	228	820	238
Ranking		27	27	18	16	19	24	22	21	14	29	23	17	5	8	26	9	2	15	20	10	6	11	11	11	1	7	4	3	25

FIGURE 12.13
House of Quality 1: customer and technical process requirements.

Quality Function Deployment—House of Quality 2
Process Components

Process Requirements	Importance	Schedule Service	Register Patient	Perform Service	Provide Imaging Results	Provide Spiritual Care	Connect to Cancer Center	Authorize Service	Process Self-Referral	Request Records/Films	Receive Referral	Perform Surgery	Process VIP Patient
Process Requirements													
Valet parking	312												
Close in parking	77												
Interior design	335			3									9
Service levels of process times	192	9	9	9	9			9		9		9	9
Design for Six Sigma	487	9	9	9	9			9		9		9	9
Concierge	642	3	3			1	1			1			9
Patient navigator	685	9	2	3	9	1	9		3	3			9
Warm transfer between central scheduling and doctor	331	9											
VIP and girlfriend service	801	9	9	9	9	9	9	9		3		3	9
Next-day results	367	9	9	9	9	3	3						
Same-day appointments for walk-ins	488	9											
Patients with no physician, referred to doctor	109								9				
Ability to seamlessly register patients in information system	333		9										
Referring physician preference cards	10				9								
Navigator database	289	9			9	3	9			1		3	9
Provide spiritual care	222					9	3						
Connect patient to cancer center	202					3	9						
Direct patient to financial assistance, if needed	193		9	9									
Ability to receive films/records	266									9			

	(value)	43,056	37,145	42,636	49,266	22,864	30,381	29,475	5,724	24,100	2,688	18,813	55,926
Provide films/records to second doctor	0									9			
Provide surgery	0											9	
View VIP status in Meditech	70												9
Same-day results for VIP and request	610			3	9								9
Same-day screening, diagnostic, other procedures of VIP, other as requested	747		3	9	9	1	1	9					9
Spiritual care hotline	105					9							
Trained and certified staff	228	9	9	9	9	3	9	9	1	3	1	9	9
Comprehensive women's services	820	9	9	9	9	9	9	9	3	9	3	9	9
Superior imaging technology	238	9	9	9	9	9	9	9	3	9		9	9
Weighting		43,056	37,145	42,636	49,266	22,864	30,381	29,475	5,724	24,100	2,688	18,813	55,926
Ranking		7	5	3	2	8	6	4	11	10	12	9	1

FIGURE 12.14
House of Quality 2: Technical process requirements and process components.

Quality Function Deployment—House of Quality 3

Metrics

Process Requirements	Importance	Days to Schedule Service	Time to Register	Time to Perform Service	Wait Time for Register	Wait Time for Service	Time to Provide Results (Internal)	Time to Provide Results (External)	Time to Connect to Spiritual Care	Percent (%) Patients Receiving Spiritual Care	Time to Connect to Cancer Center	Percent (%) Patients Connect to Cancer Center	Time to Authorize Service	Time to Provide Records/Films	Time to Perform Surgery	Percent (%) of VIP Patients Processed
Schedule service	6	9														
Register patient	8		9		9											
Perform service	10			9		9										
Provide imaging results	11						9	9								
Provide spiritual care	5								9	9						
Connect to cancer center	7										9	9				
Authorize service	9												9			
Process self-referral	2															
Request records/films	3													9		
Receive referral	1															
Perform surgery	4														9	
Process VIP patient	12															9
Weighting		54	72	90	72	90	99	99	45	45	63	63	81	27	36	108
Ranking		11	7	4	7	4	2	2	12	12	9	9	6	14	15	1

FIGURE 12.15
House of Quality 3: Process components and process metrics.

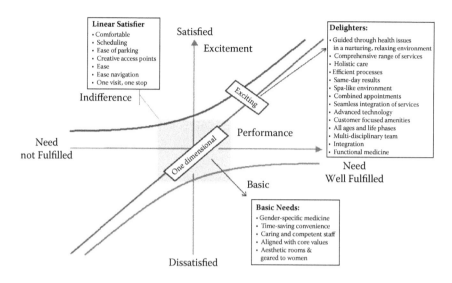

FIGURE 12.16
Kano model for Women's Center.

12.2.4.4 Kano Analysis

The Kano analysis was used to identify the exciting performance-related critical to satisfaction criteria that would provide excitement and be exciters to the Women's Center patients, and the basic factors that if missing would be dissatisfiers. The Kano analysis is shown in Figure 12.16.

12.3 Design Phase Activities

1. *Design Report*: Create a design phase report, including your findings, results, and conclusions of the Design phase.

2. *Develop New Processes*: Develop the processes based on the customer requirements.

3. *Perform Process Analysis*: Prepare a process analysis for the proposed process.

4. *Perform Waste Analysis*: Perform a waste analysis for the proposed process.

5. *Create Operational Definitions and Conceptual Model*: Create the operational definitions for the proposed metrics. A conceptual model describes the roles, information, and elements required to provide services through the processes. Develop a conceptual model for the elements of your processes.

12.3.1 Design Phase

The following is a written report of the Design phase for the Women's Center Process Design project. The Design phase of the DFSS process is focused on designing a process and the potential failures so they are reduced or eliminated with the potential to achieve a Six Sigma quality level. The main activities of this phase are as follows: identify the process elements, design the process, and identify potential risks and inefficiencies.

12.3.1.1 Develop New Processes

A business scenario method was used to develop the new processes. The following are the steps that we applied (TOGAF, 2011):

1. *Identify the business scenario objectives*: Design the new processes to meet the customer requirements and CTS characteristics identified in the VOC analysis.
2. *Identify the business scenarios to generate*: The core processes were identified for the business scenarios.
   - Schedule service
   - Register patient
   - Perform service
   - Provide results
   - Provide spiritual care
   - Connect to cancer center
   - Authorize service
   - Self-referral
   - Request films
   - Receive referral
   - Perform surgery
   - VIP (very important person) processing
   - Patient navigator processing
3. *Develop the business scenarios*, asking the following questions for each scenario:
   - What is the overall objective of the scenario?
   - What are any preconditions that must exist to start the scenario?
   - What are the inputs (information, etc.) used to start the scenario?
   - Walk through the "happy path" steps to get to the end result.

- Walk through alternate steps to get to the end result; and "not" to get to the end result (failure).
- Think of who the person would interact with along the way.
- Think of what systems the person might interact with (generally).
- Think of assumptions or needs the person might expect to require during the scenario.
- What is the final outcome, result, or output?
- What are the resources and people needed?

4. *Review the scenarios with the team and revise as necessary*

5. *Develop the process maps from the business scenarios*

The business scenario template is shown in Figure 12.17.

A sample business scenario for Provide Imaging Results is shown in Figure 12.18.

12.3.1.2 Perform Process Analysis

The process analysis defined the process inefficiencies for each proposed process (Figure 12.19).

The following gaps were identified between the current and future state process maps:

- Lack of seamless integration between the hospital medical information system and the medical group information system
- Triage for spiritual care
- Referral physician preference cards
- Concierge process
- Patient navigator and patient navigator process
- Patient navigator database and application to track patient interaction
- Track utilization of spiritual care
- Track connection to cancer center
- Defined service levels for processes
- Next-day results not meeting marketed expectations of same-day results
- Seamless online appointment scheduling
- Navigating through the Women's Center
- Detailed future state process maps need to be completed

Business Scenario Template

BUSINESS SCENARIO GENERAL INFORMATION	
Use case name	
Use case ID	
Intent	
Start (trigger) stimulus	

CUSTOMERS	
Primary customers	
Secondary customers	

PRECONDITIONS	
1.	
2.	
3.	
4.	

ASSUMPTIONS			
Number	Assumption	Date	Owner
1.			
2.			
3.			
4.			
5.			

BASIC FLOW					
Number	Step Description	Inputs	People Responsible	Outputs	Equipment, Information Technology (IT, etc.)
1.					
2.					
3.					
4.					
5.					
6.					
7.					
8.					
9.					
10.					
11.					
12.					
13.					

POSTCONDITIONS	
Course	Description
1.	Ideal	
2.	Alt-A	
3.	Alt-B	
4.	Alt-C	

ALTERNATE A					
Number	Step Description	Inputs	People Responsible	Outputs	Equipment, Information Technology (IT, etc.)
14.					
15.					
16.					
17.					
18.					
19.					
20.					
21.					
22.					
23.					
24.					
25.					
26.					

REVISION HISTORY			
Date	Version	Description	Author

FIGURE 12.17
Business scenario template.

The team developed and reviewed the process maps based on the business scenarios. A sample process map is shown in Figure 12.20. The following process maps were developed:

- Schedule
  - Imaging service
  - Medical doctor
  - Cancel or reschedule

- Register patient
  - Register imaging patient
  - Register imaging walk-in
  - Register physician patient
  - Register physician patient walk-in

BUSINESS SCENARIO GENERAL INFORMATION	
Use case name	**Provide Imaging Results**
Use case ID	004
Intent	Describes the process of providing imaging results
Start (trigger) stimulus	Patient receives imaging service

CUSTOMERS	
Primary customers	Imaging patient
Secondary customers	Physician patient

PRECONDITIONS	
1.	Patient has received an imaging service
2.	
3.	
4.	

ASSUMPTIONS	
1.	A preference card database exists with referring physician preferences of when patient navigator will provide which results to their patients
2.	Patient navigator will record discussions with patients, and call attempts when not successfully reaching the patient
3.	Patient navigator coordinates navigation of patient through Women's Center

BASIC FLOW: *Provide Imaging Results—Screening Mammography*					
Number	Step Description	Inputs	People Responsible	Outputs	Equipment (information technology, IT, etc.)
1.	Mammography technologist completes the study	Image from radiology system	Radiology technologist		
2.	Radiologist reviews case	Image from radiology system	Radiologist		Radiology system
3.	If need diagnostic screening, patient navigator notifies referring physician.		Patient navigator		
4.	Radiologist may discuss case with referring physician	Finding	Radiologist	Discussion information	
5.	Radiologist documents physician contact	Discussion information	Radiologist	Record of physician contact	
6.	Radiologist reports into reporting system		Radiologist	Radiologist report	Reporting system
7.	Reporting system sends data to medical information system	Result data after manual upload			Reporting system

8.	Radiology staff uploads reports into medical information system three times a day	Results data	Radiology staff	Results data, letters, reports faxed/mailed to ordering physician	Results system/ Medical information system
9.	Results system creates letter to patient	Result data		Patient letter	Results system
10.	Review preference card for referring physician to identify whether can call patient with results	Preference card	Patient navigator	Preferences	???
11.	Patient navigator calls patients to discuss results for those patients that they can call, for 1, 2, 3 results.	Results data, letters	Patient navigator		
12.	If patient navigator cannot reach patient, tries several times, and records attempts	Results data	Patient navigator	Record of calls	
13.	If patient needs additional procedures, patient navigator or Dr. Office will facilitate appointment. Patient navigator can call centralized scheduling with the verbal order, and do a warm transfer with the patient for demographics. Use Schedule Service Use Case	Patient info	Patient navigator	Appointment request	
14.	Patient navigator ensures that certified mail is sent and retains records for follow-up.	Patient	Patient navigator	Record of certified mail	
15.	Create full report, then upload report into reporting system, generates final report that goes into medical information system	Results data	System	Full report	Medical information system
16.	Patient receives results with letter, patient navigator, and referring physician.	Results		Results	
17.	If patient requests, radiologist talks to patient	Patient request	Radiologist		

		Postconditions		Service-Level Goals
	Course		Description	
1.	Ideal	Patient receives screening mammography results.		
2.	Alt-A	Provide imaging results—diagnostic mammography		
3.	Alt-B	Provide ultrasound results		
4.	Alt-C	Provide biopsy results		
5.	Alt-D	Provide bone density results		
6.	Alt-E	Provide results of other services		
7.	Alt-F	Patient calls patient navigator, patient navigator discusses results if permitted in referring physician preferences, or requests that patient calls referring physician		

Alternate A: Provide Imaging Results—Diagnostic Mammography					
Number	Step Description	Inputs	People Responsible	Outputs	Equipment (IT, etc.)
1.	Mammography technologist completes the study	Image from radiology system	Radiology technologist		
2.	Radiologist reviews case	Image from radiology system	Radiologist		Radiology system
3.	If positive finding, radiologist consults with physician for 4s and 5s (rare on a screening mammography).		Radiologist		
4.	Radiologist documents physician contact	Discussion information	Radiologist	Record of physician contact	
5.	Radiologist reports into reporting system		Radiologist	Radiologist report	Reporting system
6.	Penrad sends data to medical information system	Result data after manual upload			Reporting system
7.	Radiology staff uploads reports into medical information system three times a day	Results data	Radiology staff	Results data, letters, reports faxed/ mailed to ordering physician	Reporting system/ Medical information system
8.	Penrad creates letter to patient	Result data		Patient letter	Reporting system
9.	Patient navigator divides results into BIRADS 0s, 4s, 5s, and others	Results data, letters	Patient navigator	Results data, letters	
10.	Patient navigator checks preference cards for 4s and 5s, to verify for the prescribing physician whether they can call the patient with the results.	Referring Physician Preference Card	Patient navigator	Preferences	
11.	Patient navigator checks preference cards for 1, 2, 3s, to determine whether they can call patient with results	Preference card	Patient navigator	Preferences	
12.	Patient navigator calls patients to discuss results for those patients that they can call, for 1, 2, 3s.	Results data	Patient navigator		
13.	If patient navigator cannot reach patient, tries several times, and records attempts	Results data	Patient navigator	Record of calls	
14.	For 4s and 5s, patient navigator will wait a few days and ensure that the physician office and the patient received the results, and then call the patient	Patient information	Patient navigator	Patient results	

15.	If patient needs additional procedure, patient navigator or Dr. Office will facilitate appointment. Patient navigator can call centralized scheduling with the verbal order, and do a warm transfer with the patient for demographics. Use schedule service use case.	Patient info	Patient Navigator	Appointment request	
16.	Patient navigator ensures that certified mail is sent and retains records for follow-up.	Patient	Patient navigator	Record of certified mail	

Alternate B: Provide Ultrasound Results					
Number	**Step Description**	**Inputs**	**People Responsible**	**Outputs**	**Equipment (IT, etc.)**
1.	Same steps 1 to 2 as in basic flow				
3.	Radiologist dictates report into transcription system	Transcription system	Radiologist	Final reports upload automatically into Meditech and it is faxed/ mailed to ordering physician	Transcription system/ Medical information system
4.	Radiologist discusses positive findings with physician as appropriate	Findings	Radiologist	Discussion	
5.	Patient navigator reviews U.S. results	Results	Patient navigator	Results	
6.	Patient navigator provides results to referring physician	Results	Patient navigator	Results	
7.	Referring physician provides results to patient	Result, patient information	Referring physician	Results	
8.	Patient navigator reviews preference card. If referring physician preferences permits, patient navigator discusses results with patient	Results	Patient Navigator	Discussion	

Alternate C: Provide Biopsy Results					
Number	Step Description	Inputs	People Responsible	Outputs	Equipment (IT, etc.)
1.	Perform procedure biopsy-specimen sent to pathology		Radiologist/ technologist		Stereotactic unit
2.	Radiologist dictates procedure after pathology report becomes available	Transcription system	Radiologist/ patient navigator	Final report to ordering physician	Transcription system/ Radiology system
3.	Patient navigator discusses biopsy results with referring physician	Results	Patient navigator	Results	
4.	Referring physician provides biopsy results with patient	Results	Referring physician	Results discussion	
5.	Report is sent to patient in certified mail	Results	Patient navigator	Results	
6.	Patient navigator verifies results are received in certified mail and documents receipt	Results	Patient navigator	Record certified mail results	
7.	Patient navigator reviews preference card. If referring physician preferences permit, patient navigator discusses results with patient	Results	Patient navigator	Discussion	

Alternate D: Provide Bone Density Results					
Number	Step Description	Inputs	People Responsible	Outputs	Equipment (IT, etc.)
1.	Procedure is performed		Tech		Bone density unit
2..	Printed graphs sent to doctor	Radiology staff	Radiology technologist/ radiology staff		
3.	Doctor dictates final report into transcription system	Transcription system	Doctor	Final reports uploaded to Meditech automatically faxed/mailed to referring physicians	Medical information system
	Add color graphs				
4.	Returned to WC for mail distribution	Radiology staff	Radiology staff	Reports mailed to referring physicians	
5.	Final report mailed to referring physician	Report	Radiology staff	Mailed report	
6.	Referring physician provides results to patient	Report	Referring physician	Report	
7.	Patient navigator reviews preference card. If referring physician preferences permits, patient navigator discusses results with patient	Report	Patient navigator	Discussion	

Alternate E: Provide Results of Other Services					
Number	Step Description	Inputs	People Responsible	Outputs	Equipment (IT, etc.)
1.	Complete other service				
2.	Provide report of results	Result	Radiologist or other provider	Results	
3.	Provide results to referring physician	Results	Patient navigator	Results	
4.	Referring physician provides results to patient	Results	Referring physician	Results	
5.	Patient navigator reviews preference card. If referring physician preferences permits, patient navigator discusses results with patient.	Results	Patient navigator	Results	

Alternate F: Patient Navigator Discusses Results if Permitted in Referring Physician Preferences, or Requests That Patient Calls Referring Physician					
Number	Step Description	Inputs	People Responsible	Outputs	Equipment (IT, etc.)
1.	Patient calls patient navigator for results	Patient information	Patient	Patient call	
2.	Patient navigator reviews patient chart and results	Results, patient info	Patient navigator	Results	
3.	Patient navigator review referring physician preference card	Preference card	Patient Navigator	Preferences	
4.	If preferences permit, patient navigator reviews results with patient. If preferences do not permit review of results, requests for patient to call referring physician.	Preferences, results	Patient navigator	Discussion	

Revision History			
Date	Version	Description	Author
6/10/10	1.0	Draft prior to scenario building	Sandy Furterer
8/14/10	1.1	Revised based on input and review from team	Sandy Furterer

FIGURE 12.18
Business scenario provide results.

- Authorize patient
- Provide spiritual care
- Connect to cancer center
- Process self-referral
- Perform surgery
- Perform imaging service
- Perform physician service

Inefficiency
Process: Schedule Service
Having to do a warm transfer from Central Scheduling to Women's Center (WC) Medical Group physician
Process: Register Patient
If patient is not preregistered, know late in the registration process that he or she is a VIP (not prior to arriving in WC)
Should ensure all patient items are available and accurate prior to registration in WC (labs, script, ID)
Ensure only needed patient information is provided from the patient when they register an existing patient (between imaging and doctor)
Authorize Service
Multiple rework loops if authorization is not OK or correct
Having to obtain a changed order for Medicare patient
Provide Spiritual Care
Resourcing for WC needs to be identified
Connect to Cancer Center
Potential of not having preference card or not having it updated
Self-Referral
Identify if accept self-referrals
Perform Surgery
Rework getting labs, prep work, patient information, and so forth
Multiple calls getting health history; informing of copay
Getting lab results timely
Having to bill from radiology to surgery and then have radiology reimbursed

FIGURE 12.19
Potential process inefficiencies.

- Provide results
  - Screening mammogram
  - Diagnostic mammogram
  - Ultrasound results
  - Biopsy results
  - Bone density results
  - Other results
- Patient navigator
- Request records/films
- Process VIP patients

12.3.1.3 Perform Waste Analysis

The following wastes were identified for the future state processes, shown in Figure 12.21.

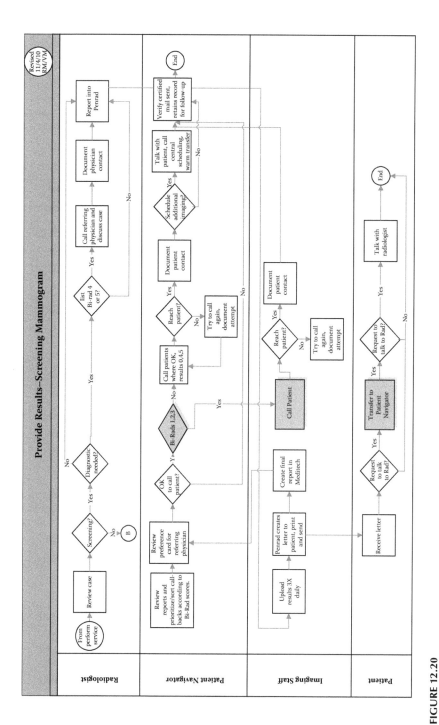

FIGURE 12.20
Sample process map: provide results—screening mammogram.

Process	Types of Waste
• Schedule	Overproduction: transfer between central scheduler and medical group Processing: duplicate information in medical information and physician system.
• Register patient	Defect: VIP patient not identified prior to visit or during visit Defect: patient forgets prescription Processing and Delay: need to call physician's office for prescription Delay: register walk-ins
• Authorize patient	Defect: wrong insurance Defect: wrong authorization Delay: verify medical necessity Processing: receive change order
• Provide spiritual care	Delay: availability of resources
• Connect to cancer center	Motion: patient going from Women's Center (WC) to cancer center Delay: due to not having physician preferences to contact patient
• Process self-referral	Delay: referral to physician
• Perform surgery	Overproduction: scheduling and patient information in surgery system
• Perform imaging service	Delay: wait times in lobby, procedure room
• Perform physician service	Delay: wait times in lobby, room
• Provide results	Delay: providing results Delay: physician providing results Delay: not having physician preferences for patient contact Delay: radiologist reading Delay: not reaching patient
• Patient navigator	Inventory: patient navigator capacity
• Request records/films	Delay: printing films
• Process VIP patients	Defect: VIP patient not identified as VIP Delay: difficulty fitting VIP in schedule

FIGURE 12.21
Waste analysis.

12.3.1.4 Create Operational Definitions and Conceptual Model

The following operational definitions were developed for the results times:

Operational definition: Time from arrival in imaging reception to exam

Defining the measure: Focus is on the time it takes to present at the radiology reception desk until procedure initiated

Purpose: We want to understand how long patients wait.

Clear way to measure the process: We will measure the time that a patient waits in the waiting area of the Radiology department. This will be measured by calculating the elapsed time from when the patient presents at the Imaging reception desk until they enter the procedure room. To baseline we will use QueVision data from 4/1 to 6/24 and subtract CheckInTime from InProcessTime.

Operational definition: Time from procedure start to procedure completion

Defining the measure: The focus is on the time it takes to complete procedures.

Purpose: We want to understand how long a procedure takes.

Clear way to measure the process: We will measure the time that a patient spends in a procedure room in the Radiology department. This will be measured by calculating the elapsed time from when the patient enters the procedure room until the procedure is complete. To baseline we will use QueVision data from 4/1 to 6/24 and subtract InProcessTime from CompletedTime:Mammo.

Operational definition: Time from procedure completion to when results are available

Defining the measure: The focus is on the time it takes to result and transmit procedure results

Purpose: We want to understand how long it takes for procedure results to be available.

Clear way to measure the process: We will measure the time that a radiologist takes to result procedures and for the techs to upload them into the medical information system for transmission. This will be measured by calculating the elapsed time from when the procedure is complete until it is available to the primary care physician (PCP).

Operational definition: Time from exam completion to review with patient

Defining the measure: The focus is on the time it takes to communicate results to patients.

The Women's Center Conceptual Model helps to identify the key elements required for the Women's Center, shown in Figure 12.22.

12.3.2 Design Phase Case Discussion

1. Design Report
    a. Review the design report and brainstorm some areas for improving the report.
    b. How did your team ensure the quality of the written report? How did you assign the work to your team members? Did you face any challenges of team members not completing their assigned tasks in a timely manner, and how did you deal with it?
    c. Did your team face difficult challenges in the Design phase? How did your team deal with conflict on your team?
    d. Did your instructor and/or Black Belt or Master Black Belt mentor help your team better learn how to apply the Design for Six Sigma tools in the Design phase, and how?
    e. Did your Design Phase Report provide a clear understanding of the root causes of the discipline process, why or why not?

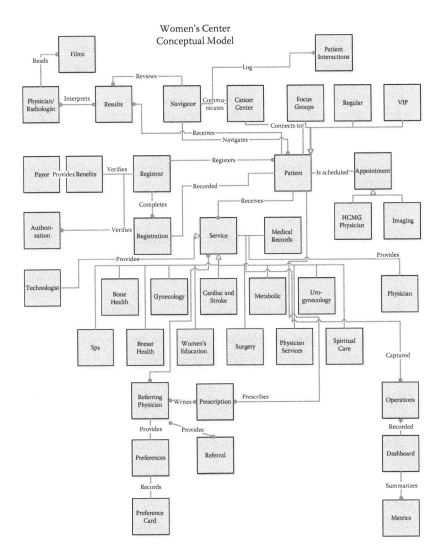

FIGURE 12.22
Women's Center conceptual model.

2. Process Map
   a. Was it difficult to create a process map for the process, and also the procedures?

3. Process Analysis
   a. Discuss how your team defined whether the activities were value-added or non-value-added activities. Was the percent of value-added activities what you would expect for this type of process and why?

4. Waste Analysis
   a. What types of waste were prevalent in this process and why?
5. Operational Definitions
   a. What other metrics could you identify and measure?
   b. Was it difficult to clearly define the operational definition?
   c. Was it difficult to identify the elements of your conceptual model?

12.4 Optimize Phase Activities

12.4.1 Optimize Phase Exercises

1. *Optimize Report*: Create an Optimize Phase report including your findings, results, and conclusions of the Optimize phase.
2. *Implementation Plan*: Develop an implementation plan for the designed process.
3. *Statistical Process Control*: Develop an example of a control chart that could be used to ensure that the process stays in control.
4. *Process Capability*: Perform a capability analysis to assess whether the process is capable of meeting the target metrics.
5. *Revised Process Map*: Revise your process map to incorporate improvements that will further enhance the process.
6. *Training plans, procedures*: Create a training plan and a detailed procedure for the new process.

12.4.2 Optimize Phase

The purpose of the Optimize phase is to pilot the new processes and assess whether they are capable of meeting the desired targets. The following activities are performed:

1. Implement pilot processes.
2. Assess process capabilities.
3. Optimize design.

12.4.2.1 Implement Pilot Process

We implemented the processes when the Women's Center opened in September 2010, and compared the process times to the baseline in the previous imaging center. A detailed work plan was developed that included the activities required to open the facility and implement the newly designed processes. The key process elements implemented were

- Hire a patient navigator to navigate the patient through the system and communicate results of procedures.
- Hire an additional radiologist to provide results more quickly.
- Unveil a new spa-like facility.
- Incorporate additional services.
- Hire a women's services focused physician.
- Train mammography technicians in customer service.
- Incorporate spiritual care resources into the center.
- Cross-train registration staff in imaging center and physician office scheduling systems.
- Hire a concierge to guide patients to their destinations.
- Have volunteers walk patients to their destinations.
- Incorporate a VIP service.
- Implement a process dashboard for metrics.
- Incorporate advanced imaging technology.

12.4.2.2 Assess Process Capabilities

The proposed value stream map for the initial mammogram screening was developed to assess process capability, shown in Figure 12.23. The process as designed will be capable of meeting the proposed wait times, procedure times, throughput time, and results time.

12.4.2.3 Optimize Design

We identified potential problems in the processes after they were piloted and created three Why-Why diagrams (Figures 12.24, 12.25, and 12.26):

- Why are there excessive wait times in the WC?
- Why do procedures take more than one business day to result?
- Why does it take a long time for the patients to receive the results?

Further improvements were designed to streamline the processes, based on staffing, results turnaround, and throughput times. The patient navigator was hired in the first quarter of 2011. She has provided great value for our patients and referring physicians. She provides sociopsychological care to our patients by providing results more timely, guiding them

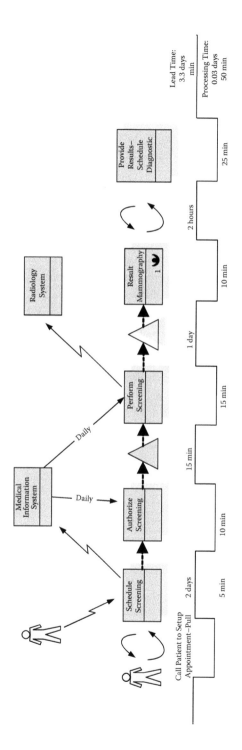

FIGURE 12.23
Proposed value stream map.

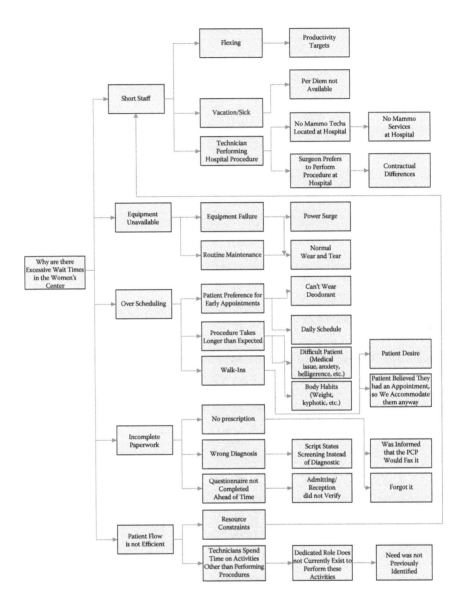

FIGURE 12.24
Why-why diagram: Why are there excessive wait times?

through the system, and connecting them to spiritual care and to the cancer center when necessary. She has been a key differentiator in the process to provide the "delighters" of the patient being guided through the health system in a nurturing, relaxing environment; comprehensive care; holistic care; efficient processes; next-day results; and seamless integration of services.

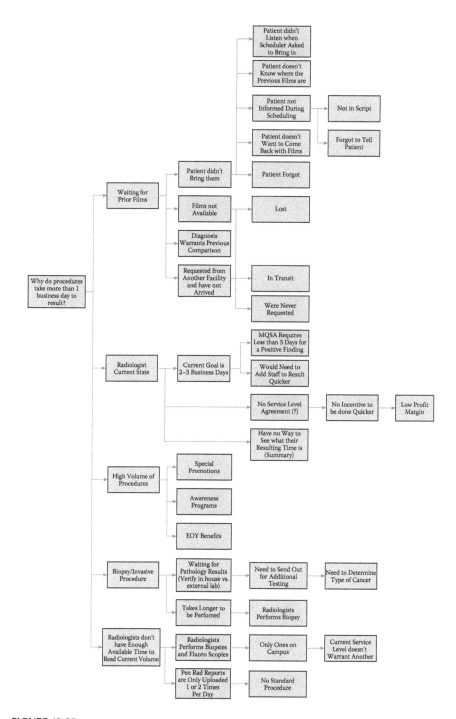

FIGURE 12.25
Why-why diagram: Why do procedures take more than one business day to result?

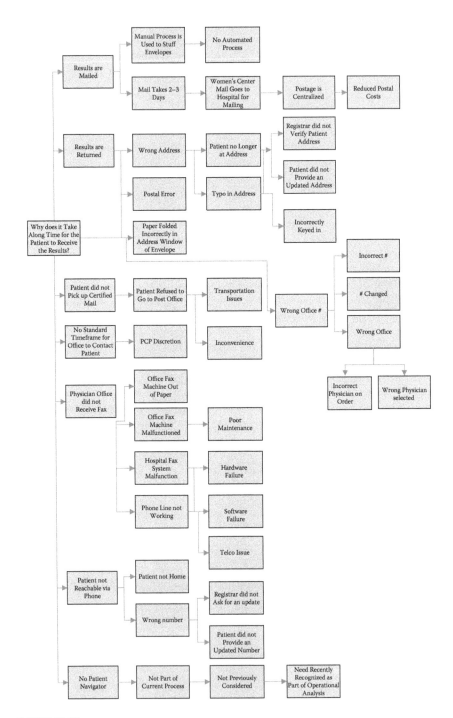

FIGURE 12.26
Why-why diagram: Why does it take a long time for the patient to receive the results?

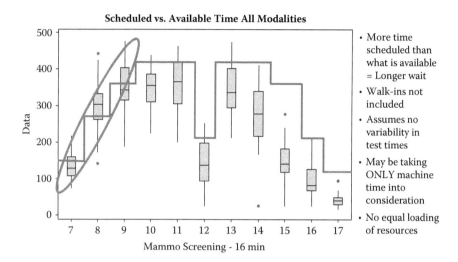

FIGURE 12.27
Takt time analysis.

A takt time throughput analysis was performed to assess whether the cycle times could achieve the needed takt time. Takt time is calculated based on the available time of scheduled resources divided by the number of scheduled procedures. Figure 12.27 shows there are times of the day when the scheduled procedures exceed the staff capacity. This analysis helped to realign the staffing to meet the patients' demand. The continuous line shows the desired takt time, and the box plots show when the average cycle time exceeds the desired takt time.

The results for the newly designed processes are shown in Figure 12.28 and compared to the older imaging facility. The waiting time in the imaging

Metric	Prior Imaging Center (Before)	Women's Center (After)	Percent (%) Improvement
Imaging waiting area	Mean = 25 minutes Standard Deviation = 30.6; Count = 3219	21 minutes	16%
Procedure time	Mean = 13 minutes; Standard deviation = 23.1; Count = 3165	14 minutes	–7%
Total throughput time	Mean = 38 minutes; Standard deviation = 36; Count = 3165	35 minutes	8%
Time to result	Mean = 2.4 days (57.2 hours); Standard deviation = 55.04 hours; Count = 6693	1 day	58%

FIGURE 12.28
Improved process times.

area improved by 16%, the procedure time increased 1 minute, on average, by 7%. The total throughput time decreased by 3 minutes or 8%, compared to the process times in the older Imaging Center, prior to opening the new Women's Center facility. The time to result tests, which was a "delighter" improved by 58% from 2.4 to 1 day.

12.4.3 Optimize Phase Case Discussion

12.4.3.1 Optimize Report

1. Review the Optimize report and brainstorm some areas for improving the report.
2. How did your team ensure the quality of the written report? How did you assign the work to your team members? Did you face any challenges of team members not completing their assigned tasks in a timely manner, and how did you deal with it?
3. Did your team face difficult challenges in the Optimize phase? How did your team deal with conflict on your team?
4. Did your instructor and/or Black Belt or Master Black Belt mentor help your team better learn how to apply the Design for Six Sigma tools in the Optimize phase, and how?
5. Did your Optimize phase report provide a clear understanding of the root causes of the discipline process? Why or why not?
6. Compare your Optimize report to the Optimize report in the book. What are the major differences between your report and the author's report?
7. How would you improve your report?

12.4.3.2 Implementation Plan

1. How must the culture be considered in an implementation plan?
2. How must the communication be considered in an implementation plan?
3. How did your Lean Six Sigma team identify the timings for when to implement your recommendations?

12.4.3.3 Process Capability

1. Why is it important to assess process capability?
2. Why is it important to ensure that your process is stable before assessing process capability?

12.5 Validate Phase Activities

1. *Dashboards/scorecards*: Create a dashboard or scorecard for tracking and controlling the process.

2. *Mistake proofing*: Create a mistake-proofing plan to prevent errors from occurring in the process.

3. *Hypothesis testing/analysis of variance (ANOVA)*: Using the data provided in the report, perform the appropriate Hypothesis test or ANOVA to compare process time metrics before and after improvement.

4. *Replication opportunities*: Identify some potential replication opportunities within or outside the division to apply the same or a similar process.

12.5.1 Validate Phase

The Validate phase validates the processes and ensures that they meet the customers' needs. The following activities are performed in the Validate phase:

1. Validate processes.
2. Assess performance, failure modes, and risks.
3. Iterate design and finalize.

12.5.1.1 Validate Processes

A dashboard was developed that could be reviewed continuously during the day, to assess the wait times, procedure times, throughput times, patients waiting for procedures, volume, and results times. A more static dashboard was created that would track performance on a monthly basis, shown in Figure 12.29. The metrics were being met for average procedure time, average wait time, and average throughput time, but were not meeting targets for percent of patients waiting more than 45 minutes and percent of mammography patients being contacted by the navigator within 5 days of service. We continued to adjust staffing and improve the processes to better meet the targets.

12.5.1.2 Assess Performance, Failure Modes, and Risks

We assessed the potential failure modes in the processes and identified the potential risks:

1. Focusing on productivity as the key metric impacts our ability to meet the quality indicators of patient wait and throughput times.

Women's Center Mammography Throughput KPI Dashboard

	Target	January	February	March
Quality Indicators				
Average mammogram procedure time (screening and diagnostic)	<= 18 min.			
Average mammogram wait time	<= 25 min.			
Average mammogram throughput time	<= 45 min.			
Percent (%) of patient waiting >45 minutes	5%			
Percent (%) mammogram patients contacted by navigator within 5 days of date of service	100%			
Total patients				

	Green = meeting target
	Red = not meeting target

FIGURE 12.29
Dashboard.

2. The patient navigator continues to meet with referring physicians to describe the patient navigator services available, and complete physician preference cards that identify when it is appropriate for the patient navigator to share the patient results directly with the patients and the time frame to do so.

3. Identifying the VIP members has been problematic with an information systems issue. This issue has been logged for the Information Technology Help Desk.

4. There are several outstanding action items that have not yet been resolved and need the executive team to embrace and resolve the more political issues related to growth, volume, and physician relationships.

12.5.1.3 Iterate Design and Finalize

The new design has been stabilized but will be continually improved and monitored.

12.6 Conclusions

The Design for Six Sigma methodology was extremely successful in capturing the voice of the customer and translating the customer requirements into the process requirements and for designing a process that delights the customers of the Comprehensive Women's Center. The opening of the new Women's Center has surpassed the organization's and the customers' expectations. The facility provides comprehensive women's services in a nurturing and spa-like environment. The future looks bright for the next several phases of the center.

12.6.1 Validate Phase Case Discussion

1. Validate Report
   a. Review the Validate Report and brainstorm some areas for improving the report.
   b. How did your team ensure the quality of the written report? How did you assign the work to your team members? Did you face any challenges of team members not completing their assigned tasks in a timely manner, and how did you deal with it?
   c. Did your team face difficult challenges in the Validate phase? How did your team deal with conflict on your team?
   d. Did your instructor and/or Black Belt or Master Black Belt mentor help your team better learn how to apply the Design for Six Sigma tools in the Validate phase, and how?
   e. Compare your Validate report to the Validate report in the book. What are the major differences between your report and the author's report?
   f. How would you improve your report?
2. Dashboards/Scorecards
   a. How would your dashboard differ if it was going to be used to present to the executive management versus the departmental management?
3. Mistake Proofing
   a. How well did your team assess the mistake-proofing ideas to prevent errors?
4. Hypothesis Tests, Design of Experiments
   a. How did you assess the improvement for the CTS?
5. Replication Opportunities
   a. How did your team identify additional replication opportunities for the process within and outside the information system division?

References

Engelman, K., Cizik, A., Ellerbeck, E., Women's Satisfaction with Their Mammography Experience: Results of a Qualitative Study, *Women & Health*, 42(4), 2005.

TOGAF, 26. Business Scenarios, http://pubs.opengroup.org/architecture/togaf9-doc/arch/, 2011.

13

Project Charter Review Process Design— A Design for Six Sigma Case Study

Sandy Furterer

CONTENTS

13.1 Project Overview

The Information System Division of a major Fortune 500 corporation develops applications to support the business. The division had been reviewing and approving the projects in a cross-divisional weekly meeting with the senior executives. The project charter is developed by the application development team working with the business to understand the scope of the proposed project. The project charter includes a description of the business opportunity, identification of the customers and stakeholders, the goals and objectives of the project, as well as the metrics that assess the successful completion of the project. The project charter also includes identification of the potential risks that could prevent the project from being successfully completed, and the assumptions that are assumed to be true. An initial estimate of the resources and project costs and the hard and soft benefits for doing the project are also assessed. The hard benefits identify financial savings that impact the financial statements, while the soft benefits include cost avoidance and intangible benefits to the business for doing the project. The customer signatures signifying buy-in to the project are also included on the project charter. The division's Program Management Office (PMO) provides project management standards, guidance, and training to the division. They have recently decentralized the project charter approval process to the senior vice presidents' (SVP) areas. The review and approval of projects had been performed at a divisional level, looking only at projects that were greater than a thousand hours of effort. If projects were under 1000 hours of total effort, they were reviewed by VPs, but not across the SVP area. The goal was to get more visibility of all projects across the entire SVP area. The approval from the customer will be attained, and then the information system division SVP area will review the project charter to identify any cross-area conflicts or overlap and ensure that resources are available to work on the project.

The Process and Metrics (P&M) team in the SVP's area has been assigned the responsibility of designing a new Area Council review process to assess the quality of the project charter, and incorporate appropriate metrics to baseline and encourage continuous process improvement. The divisional standards should be maintained to ensure consistency and repeatability of the project chartering process. The stakeholders of the new process include the management team, who will review and approve the projects within the information system division; the Process and Metrics team, who will assess the quality of the project charters and execute the Area Council review process; the PMO, who provides the divisional standards for reviewing the project charters; the project leaders, who create the project charters; and the business, for whom the project charters are developed.

13.2 Identify Phase Exercises

It is recommended that the students work in project teams of four to six students throughout the Design for Six Sigma (DFSS) case study.

1. *Identify Phase Written Report*: Prepare a written report from the case study exercises that describes the Identify phase activities and key findings.

2. *Design for Six Sigma Project Charter*: Use the information provided in the "Project Overview" section above, in addition to the project charter format to develop a project charter for the Design for Six Sigma project.

3. *Stakeholder Analysis*: Use the information provided in the "Project Overview" section above, in addition to the stakeholder analysis format to develop a stakeholder analysis, including stakeholder analysis roles and impact definition, and stakeholder resistance to change.

4. *Team Ground Rules and Roles*: Develop the project team's ground rules and team members' roles.

5. *Project Plan and Responsibilities Matrix*: Develop your team's project plan for the DMAIC project. Develop a responsibilities matrix to identify the team members who will be responsible for completing each of the project activities.

6. *Identify Phase Presentation*: Prepare a presentation (PowerPoint) from the case study exercises that provides a short (10 to 15 minutes) oral presentation of the Identify phase deliverables and findings.

13.2.1 Identify Phase

13.2.1.1 Identify Phase Report

Following is a written report of the Identify phase for the project charter review process design project, including the key deliverables developed as part of the prior exercises. The main purpose of the Identify phase is to understand the opportunity and business that needs a new process to be designed, and to develop a project charter and appropriate scope to design the process. The main activities in the Identify phase are to develop project charter, perform stakeholder analysis, and develop project plan.

13.2.1.2 Design for Six Sigma Project Charter

The Process and Metrics (P&M) team in the Information System Division's Senior Vice President's (SVP) area has been assigned the responsibility of designing a new Area Council review process to assess the quality of the project charter, and incorporate appropriate metrics to baseline and encourage continuous process improvement. The divisional standards should be maintained to ensure consistency and repeatability of the project chartering process. The stakeholders of the new process include the management team, who will review and approve the projects within the information system division; the Process and Metrics team, who will assess the quality of the project charters and execute the Area Council review process; the PMO, who provides the divisional standards for reviewing the project charters; the project leaders, who create the project charters; and the business, for whom the project charters are developed.

Following are the sections that compose the project charter, which defines the problem to be investigated. The project charter is shown in Figure 13.1.

The Information System Division develops applications to support the business. The division's Program Management (PM) office members along with the division's management had been reviewing and approving the projects in a cross-divisional weekly meeting with the senior executives. The project charter is developed by the application development team working with the business to understand the scope of the proposed project. The project charter includes a description of the business opportunity, identification of the customers and stakeholders, the goals and objectives of the project, as well as the metrics that assess the successful completion of the project. The project charter also includes identification of the potential risks that could prevent the project from being successfully completed, and the assumptions that are assumed to be true. An initial estimate of the resources and project costs and the hard and soft benefits for doing the project are also assessed. The hard benefits identify financial savings that impact the financial statements, while the soft benefits include cost avoidance and intangible benefits to the business for doing the project. The customer signatures signifying buy-in to the project are also included on the project charter. The division's Program Management Office (PMO) provides project management standards, guidance, and training to the division. They have recently decentralized the project charter approval process to the senior vice presidents' (SVP) areas. The approval from the customer will be attained, and then the information system division SVP's area will review the project charter to

Project Name: Project Charter Review Process Design
Problem Statement: To design a process for the area to review the project charters to determine if the project should move forward to the next phase. A project charter is the project initiation document that identifies a need in the business to perform information systems work.
Customer/Stakeholders: The Program Management Office (PMO) has recently decentralized the review of project charters to the areas, resulting in the need for creating a process to review the project charters to ensure that they are providing value to the business, and communicating the type of information needed to identify risks and manage projects and resources at an area level.
What Is Important to These Customers—Critical to Satisfaction (CTS): All necessary fields are completed; provide accurate information to make decisions; review is timely. Obtain approval to continue with the project.
Goal of the Project: To provide a process that provides a timely and complete review and decision to continue (or not) with the project.
Scope Statement: This process includes the review of the project charters at an area level. Includes project review of the project charter, provides review of the format and content of the project charter, and provides approval of the project charter at appropriate management levels. Link this process to the quality goals of the organization. This process is just for the identified area.
Financial and Other Benefit(s): Consistent process, visibility of projects across area to identify overlap and resource sharing.
Project Deliverables: Project charter review process; scorecard and metrics with baseline and target goals, and appropriate visibility of reporting requirements.
Potential Risks: Being perceived as a bureaucratic instead of value-added process; acceptance and adherence of process of area; timeliness of review.

FIGURE 13.1
Project charter.

identify any cross-area conflicts or overlap and ensure resources are available to work on the project.

Project Name: Project Charter Review Process Design

Problem Statement: To design a process for the area to review the project charters and determine if the project should move forward to the next phase. Project charters are the project initiation document that identifies a need in the business to perform information systems work.

Customers/Stakeholders: The primary stakeholders are the management team, who will review and approve the project charters, the project leaders, who will develop the project charters, and the Process and Metrics team, who will execute the process. The secondary stakeholders are the customers, for whom the application development teams are developing applications, and the Program Management Office, who develops and ensures divisional standards are followed.

What is Important to These Customers (Critical to Satisfaction, CTS): The management team wants a simple and timely process that provides visibility of the status of the projects that enable the teams to meet the business' information system needs. The project leaders want their projects approved and want a timely and manageable process. The Process and Metrics team wants to implement a simple and measureable process that is of high quality. The business customers want the desired functionality to be delivered in a timely manner. The Program Management Office wants the standards to be followed in a consistent and repeatable manner.

Goal of the Project: To provide a process that provides a timely and complete review and decision to continue (or not) with the project.

Scope Statement: This process includes the review of the project charters at an area level. It includes project review of the project charter, review of the format and content of the project charter, and approval of the project charter at appropriate management levels. It should link this process to the quality goals of the organization. This process is just for the identified area.

Projected Financial and Other Benefits: Consistent process, visibility of projects across area to identify overlap and resource sharing.

Risk Management Matrix: The risk management matrix is shown in Figure 13.2. The main risks are not having time to get buy-in from the major stakeholders; communication of the new process may not be complete and of high quality; need to consider needed training and rollout; being considered as a bureaucratic rather than a value-added process; and timeliness of the review.

Potential Risks	Probability of Risk (H/M/L)	Impact of Risk (H/M/L)	Risk Mitigation Strategy
Not having time to get buy-in from key stakeholders	H	H	Create a simple process that can be enhanced.
Identify key stakeholders, and get input quickly.			
Communication of the new process is not complete and high quality	L	M	Identify key stakeholders and create communication and change strategy.
Need to consider needed training and roll out	H	H	Create training and roll-out strategy.
Being perceived as a bureaucratic instead of value-added process	H	H	Alignment with business and project strategies, with value clearly defined. Projects that are not resourced or aligned should not move forward.
Acceptance and adherence of process to area	H	M	Develop change management strategy.
Timeliness of review. Program reviews every 2 weeks, instead of weekly. Potential maximum impact to project is 3 weeks.	H	H	Clearly document the process and procedures to help ensure better planning. Contingency process steps may be needed.

FIGURE 13.2
Project risk matrix.

Project Resources:

Master Black Belt Mentor

Project Team Members

Project Deliverables: Project charter review process; scorecard and metrics with baseline and target goals; and appropriate visibility of reporting requirements

The business case for this project is that divisional management has made a decision to review information system project charters at the senior vice president level to provide visibility at the area level. This created an immediate need to design an Area Council review process within the area to ensure consistency across the division and enable improvement and visibility within the area.

13.2.1.3 Customers/Stakeholder Analysis

The Program Management Office (PMO) recently decentralized the review of project charters to the areas, resulting in the need for creating a process to review the project charters to ensure they are providing value to the business, and communicating the type of information needed to identify risks and manage projects and resources at an area level. The newly formed Process and Metrics team was assigned the task to design a new Area Council review process. The team decided that they will use the Design for Six Sigma tools and IDDOV (Identify-Define-Design-Optimize-Validate) methodology to ensure a fact-based process is used to design the new process, and to ensure that appropriate measures are incorporated into the process. The primary stakeholders are the management team, who will review and approve the project charters, the project leaders, who will develop the project charters, and the Process and Metrics team, who will execute the process. The secondary stakeholders are the customers, for whom the application development teams are developing applications, and the Program Management Office, who develops and ensures divisional standards are followed.

Figure 13.3 shows the primary and secondary stakeholders, and their major concerns. Note that "+" represents a positive impact or potential improvement, while "–" represents a potential negative impact to the project.

Figure 13.4 shows the commitment level of each major stakeholder group at the beginning of the project.

13.2.1.4 Team Ground Rules

The team adhered to the following ground rules related to working together on the team.

- Be respectful to team members.
- Be open minded, share ideas freely.

Stakeholders	Who Are They?	Potential Impact or Concerns	+/–
Management team	Inform system division management who monitor projects	• Simple process • Timely process • Project visibility • Meet business' needs	+ + + +
Project leaders	Leaders of information system projects	• Approval to continue with the project • Timely process • Manageable process	+ + + –
Process and Metrics team	Responsible for improving the internal application development life cycle processes, and providing metrics to ensure quality and timeliness of project deliverables.	• Simple process • Measurable process • High-quality process	– + +
Business customers	Internal customers who are provided information systems to meet their business needs.	• Deliver needed functionality • Delivery in a timely manner	+ +
Program Management Office	Division Program Management Office who provides application development life cycle standards, training, and mentoring.	• Consistent process is followed	+

FIGURE 13.3
Stakeholder analysis definition.

- Provide service to each other, with focus on customers and stakeholders.
- Provide excellence to the team.
- Respect differences.
- Be supportive rather than judgmental.
- Be open to new concepts and to concepts presented in new ways. Keep an open mind. Appreciate other's points of view.
- Share your knowledge, experience, time, and talents.

Stakeholders	Strongly Against	Moderate Against	Neutral	Moderate Support	Strongly Support
Management team					XO
Project leaders		X			O
Process and Metrics team					XO
Business customers			XO		
Program Management Office			X		O
Notes: X, at start of project; O, by end of project					

FIGURE 13.4
Stakeholder commitment scale.

Activity Number	Phase/Activity	Duration	Predecessor	Resources
1.0	Identify			
1.1	Develop project charter	1 day		Team
1.2	Perform stakeholder analysis	2 days	1.1	Team
1.3	Develop project plan	2 days	1.2	Team
2.0	Define		1.0	
2.1	Collect voice of customer (VOC)	1 day		Team
2.2	Identify critical to satisfaction (CTS) measures and targets	14 days	2.1	Team
2.3	Translate VOC into technical requirements	14 days	2.2	Team
2.4	Identify CTS measures and targets	2 days	2.3	Team
3.0	Design		2.0	
3.1	Identify process elements	5 day		Team
3.2	Design process	1 days	3.1	Team
3.3	Identify potential risks and inefficiencies	3 days	3.2	Team
4.0	Optimize		3.0	
4.1	Implement process	60 days		Team
4.2	Assess process capabilities	5 days	4.1	Team
4.3	Optimize design	5 days	4.2	Team
5.0	Validate		4.0	
5.1	Validate process	30 days		Team
5.2	Assess performance, failure modes, and risks	5 days	5.1	Team
5.3	Iterate design and finalize	½ day	5.2	

FIGURE 13.5
Project plan.

13.2.1.5 *Project Plan and Responsibilities Matrix*

The detailed project plan is shown in Figure 13.5, with tasks to be completed, due date, deliverables, and resources. It includes the person or people responsible for each activity.

13.2.1.6 *Identify Phase Presentation*

The Identify phase presentation summarizing the written Identify phase can be developed by the project team.

13.2.2 Identify Phase Case Discussion Questions

13.2.2.1 *Identify Phase Written Report*

1. How did your team ensure the quality of the written report? How did you assign the work to your team members? Did you face any challenges of team members not completing their assigned tasks in a timely manner, and how did you deal with it?
2. Did your team face difficult challenges in the Identify phase? How did your team deal with conflict on your team?

3. Did your instructor and/or Black Belt or Master Black Belt mentor help your team better learn how to apply the Design for Six Sigma tools, and how?

4. Did your Identify phase report provide a clear vision of the project? Why or why not?

5. How could you improve your Identify phase report based on the Identify phase report given in the book? How could you improve the Identify phase report in the book?

13.2.2.2 Design for Six Sigma Project Charter

Review the project charter presented in the Identify phase report.

1. A problem statement should include a view of what is going on in the business, and when it is occurring. The problem statement should provide data to quantify the problem. Does the problem statement in the Identify phase written report provide a clear picture of the business problem? Rewrite the problem statement to improve it.

2. The goal statement should describe the project team's objective and be quantifiable, if possible. Rewrite the Identify phase goal statement to improve it.

3. Did your project charter's scope differ from the example provided? How did you assess what was a reasonable scope for your project?

13.2.2.3 Stakeholder Analysis

Review the stakeholder analysis in the Identify phase report.

1. Is it necessary to identify the large number of stakeholders as in the example case study?

2. Is it helpful to group the stakeholders into primary and secondary stakeholders? Describe the difference between the primary and secondary stakeholder groups.

13.2.2.4 Team Ground Rules and Roles

Discuss how your team developed your team's ground rules. How did you reach consensus on the team's ground rules?

13.2.2.5 Project Plan and Responsibilities Matrix

Discuss how your team developed their project plan and how they assigned resources to the tasks. How did the team determine estimated durations for the work activities?

13.2.2.6 Identify Phase Presentation

1. How did your team decide how many slides/pages to include in your presentation?
2. How did your team decide upon the level of detail to include in your presentation?

13.3 Define Phase

13.3.1 Define Phase Exercises

Define Report: Create a Define phase report, including your findings, results, and conclusions of the Measure phase.

Data Collection Plan and Voice of Customer: Develop a data collection plan for collecting voice of customer and process information to assess the critical to satisfaction criteria for the project.

Critical to Satisfaction Summary: Brainstorm ideas to summarize the proposed critical to satisfaction criteria and prepare a critical to satisfaction summary and targets.

Quality Function Deployment (QFD): Develop a QFD House of Quality to identify and map the customer requirements to the technical requirements of the process.

Define Phase Presentation: Prepare a presentation (PowerPoint) from the case study exercises that provides a short (10 to 15 minutes) oral presentation of the Define phase deliverables and findings.

13.3.1.1 Define Phase Written Report

13.3.1.1.1 Define Report

Following is a written report of the Define phase for the project charter review process design project, including the key deliverables developed as part of the prior exercises. The Define phase of the IDDOV process is designed to gain information on the voice of the customer (VOC) to understand the needs of the customers and begin translating those customer requirements into the processes' technical elements. The main activities of this phase are to collect VOC; identify CTS measures and targets; and translate VOC into technical requirements.

13.3.1.1.2 Data Collection Plan and Voice of Customer

The data collection plan is shown in Figure 13.6. It summarizes the potential metrics and how we would collect data to measure the metrics. This will

Critical to Satisfaction (CTS)	Metric	Data Collection Mechanism (Survey, Interview, Focus Group, etc.)	Analysis Mechanism (Statistics, Statistical Tests, etc.)	Sampling Plan (Sample Size, Sample Frequency)	Sampling Instructions (Who, Where, When, How)
Timely process	Area Council review is held first and third Tuesday of month	Track schedule of reviews	Counts of reviews	All reviews	None
High-quality process with metrics	Content quality percentage	Scorecard with content quality criteria and score; stakeholder interviews	Percentage received against grading criteria; control chart	All project charters for each review within senior vice presidents' (SVP) area	See scorecard procedures
	Format quality percentage	Scorecard with format criteria and score, stakeholder interviews	Percentage received against grading criteria; control chart	All project charters for each review within SVP area	See scorecard procedures
Accurate information	Content quality percentage	Scorecard with content quality criteria and score	Percentage received against grading criteria; control chart	All project charters for each review within SVP area	See scorecard procedures
Ability to make decisions, go/no go on projects	Percent projects decided on in each meeting	Agenda approval record, stakeholder interviews	Percentage	Each Area Council project review meeting	See review procedures
Visibility to program/ project relationships	Count of projects related to programs	Scorecard item on format scorecard, stakeholder interviews	Count	All projects reviewed	See scorecard procedures

FIGURE 13.6
Data collection plan.

include information on a proposed process and VOC information. The VOC data collection consisted of interviewing the stakeholders to understand what is critical to their satisfaction for the new process, as well as harvesting information on similar processes, and data related to initial design thoughts for the process.

The senior vice president (SVP), the vice presidents (VP), the directors, program and project leaders, the Process and Metrics team, the enterprise and solution architects, who provide cross-area planning of information system blueprints and roadmaps, and the PMO were all interviewed.

The management team (SVP, VPS, and directors) wanted to ensure the programs and projects are on budget, have resources, and have key sponsors. They also wanted to be able to decide whether they should do the project work or not, and understand how the programs and projects affect other teams within the area and the division. They want to be able to have visibility and knowledge when there is a problem or issue with the project which potentially puts the project at risk of successful completion. The management team wants to be able to understand the project priorities and have the visibility to know if they are working on the right priorities, as well as have a way to periodically review the work being performed in the area.

Some of the questions that they would ask when reviewing projects are as follows:

- What is the scope of processes in the project?
- Should we buy versus build?
- What is the impact to the business?
- Should we outsource any part of the development work?
- Do we have engagement from the business areas?
- What business resources will be required?
- Is there an existing process that they are enhancing?
- Is infrastructure needed?
- Do we have the resources necessary to do the work, or what must be reprioritized to be able to do this work?

The program and project leaders' concerns and critical drivers were to provide resource allocation and management across the programs and projects. The Process and Metrics team, who is responsible for the Area Council review process, wants a simple process that is metrics based and encourages continuous process improvement for initiating new projects and ensures that there is customer/stakeholder buy-in to the new process.

The enterprise architects were concerned about the ability to be able to see program and project dependencies and assess the impact of adding projects to the business and the information system division. They wanted to provide visibility of program and project changes and periodic updates to the programs and projects. The architects also wanted to ensure appropriate resource management.

The PMO wants a review that supports and aligns with the divisional standards and the information system development life cycle.

To measure the timely process, we would track that the Area Council review process is held every first and third Tuesday of the month. To help meet the CTS for having a high-quality process with metrics and accurate project information, we built a scorecard that would assess both format and quality of the content on the project charter. The format scorecard criteria would ensure that all of the required fields are completed. The content scorecard criteria would assess the quality of the information in the fields against the project charter standard documentation. To assess the ability to make go/no go decisions on the projects, the percent of projects reviewed and approved or deferred will be tracked at the review meetings. The visibility of projects related to programs can be tracked by counting the number of projects that have a program identified with it.

13.3.1.1.3 Critical to Satisfaction Summary

The VOC provided insight into the CTS criteria for the project, summarized below:

- Timely process
- High-quality process with metrics
- Accurate information
- Ability to make decisions, go/no go on projects
- Visibility to program/project relationships

It is important to the project leaders, application development teams, and management that the review process provides a timely review and approval of projects so the teams can get started working on the information systems projects. It is also important to have a high-quality project initiation process and that the metrics designed enable continuous process improvement and provide project charters that are well scoped. The new process will also need to enable the ability to make decisions on whether to approve the projects or not, providing information on the business opportunity, goals, and objectives of the projects. The process should also provide visibility of the projects and resources required across the area.

13.3.1.1.4 Quality Function Deployment (QFD)

The QFD House of Quality was used to ensure alignment between customer and stakeholder needs represented by the CTS criteria and the technical requirements of the process design. After collecting the VOC information that allowed insight into the CTS criteria summarized across all of the stakeholders, the Process and Metrics team brainstormed the critical elements to be designed into the new process (technical requirements). The House of Quality is shown in Figure 13.7. The Process and Metrics team assessed the strength of relationship between the CTS criteria and the design criteria. An importance rating was assigned to each of the CTS, which was then multiplied by the relationship ratings, to derive a relative weighting of the technical requirements. A Pareto chart is shown in Figure 13.8. An Area Council SharePoint (Microsoft intranet Web site software) would be a way to provide a workflow and facilitate the review process. Executive approval is another critical element that should be designed into the new process. Without it there is little chance for the organization to see the value of the reviews, if the management team is not on board. Detailed procedures and a process map will provide clear definition of the process and can be used as a training guide, along with workshops to train on the process. A scorecard with definitive

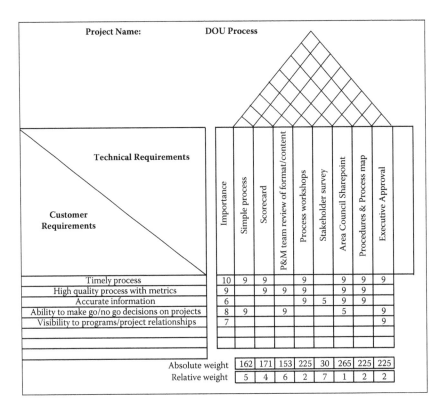

FIGURE 13.7
Quality function deployment House of Quality.

criteria for assessing the format and quality of the project charter content is the next most important design criteria, followed by the need for the Process and Metrics team to review the project charters. This will ensure consistency of the measurement process. A stakeholder survey is the last design criteria that could be used to validate the new process.

13.3.1.1.5 Define Phase Presentation

The Define phase presentation summarizing the written Define phase presentation can be developed by the project team.

13.3.2 Define Phase Case Discussion

13.3.2.1 Define Report

1. Review the Define report and brainstorm some areas for improving the report.
2. How did your team ensure the quality of the written report? How did you assign the work to your team members? Did you face any

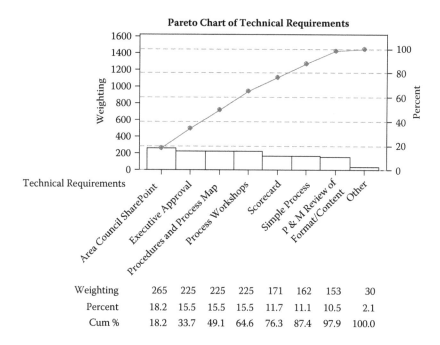

Pareto Chart of Technical Requirements

Technical Requirements	Area Council SharePoint	Executive Approval	Procedures and Process Map	Process Workshops	Scorecard	Simple Process	P & M Review of Format/Content	Other
Weighting	265	225	225	225	171	162	153	30
Percent	18.2	15.5	15.5	15.5	11.7	11.1	10.5	2.1
Cum %	18.2	33.7	49.1	64.6	76.3	87.4	97.9	100.0

FIGURE 13.8
Pareto chart of prioritized technical requirements.

challenges of team members not completing their assigned tasks in a timely manner, and how did you deal with it?

3. Did your team face difficult challenges in the Define phase? How did your team deal with conflict on your team?

4. Did your instructor and/or Black Belt or Master Black Belt mentor help your team better learn how to apply the Design for Six Sigma tools in the Define phase? How?

5. Did your Define phase report provide a clear understanding of the voice of the customer (VOC)? Why or why not?

13.3.2.2 Critical to Satisfaction Summary

1. How did you derive the CTS criteria, and how would you ensure that they represent customer and stakeholder needs?

13.3.2.3 Data Collection Plan

1. What would you perceive to be some of the difficulties of collecting VOC information in an interview format?

2. What other ways could you collect the VOC information for this project?

13.3.2.4 Quality Function Deployment

1. Why is it important to prioritize the CTS before developing the relationships between the CTS and the technical requirements?
2. Discuss how the Pareto chart provides the priority for the technical requirements.

13.3.2.5 Define Phase Presentation

1 How did your team decide how many slides/pages to include in your presentation?
2 How did your team decide upon the level of detail to include in your presentation?

13.4 Design Phase

13.4.1 Design Phase Exercises

13.4.1.1 Design Report

Create a Design phase report, including your findings, results, and conclusions of the Design phase.

Process Map: Develop a process map for the process.

Failure Mode and Effect Analysis: Create a failure mode and effect analysis, brainstorming potential failures in the project charter review process.

Process Analysis: Prepare a process analysis for the proposed process.

Waste Analysis: Perform a waste analysis for the proposed process.

Operational Definitions: Develop metrics and operational definitions that relate to the CTS for the new process.

Design Phase Presentation: Prepare a presentation (PowerPoint) from the case study exercises that provides a short (10 to 15 minutes) oral presentation of the Design phase deliverables and findings.

Design Report: Following is a written report of the Design phase for the project charter review process design, including the key deliverables developed as part of the prior exercises.

The Design phase of the DFSS process is focused on designing a process and the potential failures so they are reduced or eliminated with the potential to achieve a Six Sigma quality level. The main activities of this phase are as follows: identify process elements, design process, and identify potential risks and inefficiencies.

13.4.1.2 Process Map

The team developed the critical elements to be incorporated into the process as follows:

- Management commitment and review
- Metrics to encourage continuous improvement
- Area review meetings on the first and third Tuesdays
- Development team and VP area reviews of project charters before going to the Area Council
- SharePoint used to manage Area Council workflow and agenda
- Criteria set to review certain projects across the entire division
- Skills need to be transferred to project leads to develop high-quality project charters.
- Process needs to be simple and based on voice of customer input.

The team designed the new process using the VOC information and the process elements as a guide. The process map is shown in Figure 13.9. A description of the process follows.

13.4.1.2.1 Review Project Charter, Enter into Area Council (AC) SharePoint, Enter Scorecard

Owners: Development Team

Purpose: For the development team and the Area Director to review the project charter and ensure the completeness and content is of high quality

Steps:

1. The development team will review their project charter within their team/director area. The initiation scorecard can be used as a guide for the format and content quality levels.
2. The project leader should enter the project information in the SharePoint, with a "Status Initiation" of "pending."
3. The development team should complete the initiation scorecard via the Area Council SharePoint.

13.4.1.2.2 Approve?

Owners: Development team, manager/senior manager, director (as appropriate)

Purpose: To approve the project charter. This approval includes the format, content, and that the project charter addresses the business needs to be included in the scope of the project charter effort.

358   *Design for Six Sigma in Product and Service Development*

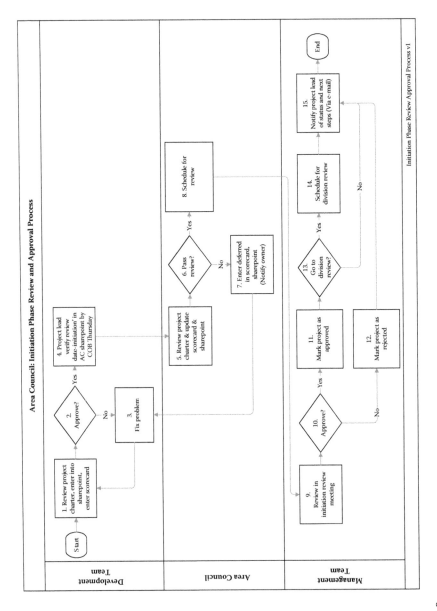

FIGURE 13.9

Process map.

Step:

1. Each director area will define their approval process.

13.4.1.2.3 Fix Problem

Owners: Development Team

Purpose: To correct any issues identified in the development team and director-level review.

Steps:

1. The development team or owner will fix any issues that are identified in the development team and director review of the project charter.
2. And resubmit for development team and director approval.

13.4.1.2.4 Project Leader Verify Review Date–Initiation in Area Council SharePoint by COB Thursday

Owners: Project leader (or development team designated owner)

Purpose: To notify Area Council that the project charter is ready for the Area Council format and content review

Step:

1. Once the project charter is approved by the development team and director, the project leader will verify the date for the "Review Date—Initiation" in the Area Council SharePoint. This "Review Date—Initiation" should correspond to the supply chain systems area program review dates (currently scheduled as the first and third Tuesday of the month.)

13.4.1.2.5 Review Project Charter and Update Scorecard and SharePoint

Owners: Area Council

Purpose: To ensure that the format and content are complete

Steps:

1. The Area Council will review the project charter for format and content, using the project charter scorecard.
2. The Area Council will update the scorecard, and communicate back to the development team as appropriate.

13.4.1.2.6 Pass review?

Owners: Area Council

Purpose: To determine if the project charter passes the scorecard criteria

Step:

    1. Make decision on initiation scorecard criteria: pass or deferred-pending changes

13.4.1.2.7 Enter Deferred in Scorecard, SharePoint (Notify Owner)

Owners: Area Council

Purpose: To notify the project leader of the development team that there are issues with the project charter to correct

Steps:

    1. Enter the decision "deferred pending changes" in the scorecard and SharePoint.

    2. SharePoint notifies the project leader of the reject and issues.

13.4.1.2.8 Schedule for Area Council (Notify Project Lead to Complete Project Charter Action Item)

Owners: Area Council

Purpose: To notify the development team and director that the project charter is scheduled for the Area Council review and the date that it is scheduled.

Steps:

1. Contact the owner (and director) that the project charter is ready to be presented at the Area Council Review, and the date that it is scheduled.

2. Add the project charter to the Area Council (program review) agenda.

3. The project lead should complete the action item in Clarity.

13.4.1.2.9 Review in Area Council

Owners: Management team

Purpose: To review the project charter and approve or reject the project charter as a project to commence further work

Steps:

1. The Area Council is to be held the first and third Tuesday of each month from 9 to 10 a.m. in the conference room.

2. The project's director (or designee) for the development team will present the project charter at the Area Council review meeting.

13.4.1.2.10 Approve?

Owners: Management team.

Purpose: To approve or reject the project charter as a project to commence further work

Step:

1. A decision will be made to "approve" or "reject" the project.

13.4.1.2.11 Mark Project as Approved in SharePoint and Clarity

Owners: Management team

Purpose: To update the Area Council SharePoint and to notify the development team's project leader of the decision.

Steps:

1. The decision will be entered into the Area Council SharePoint (during the program review meeting) and the project leader will be notified.
2. If project was approved, then vice president can approve the project in Clarity.

13.4.1.2.12 Mark Project as Rejected in SharePoint and Clarity

Owners: Management team

Purpose: To mark the project as rejected in SharePoint and Clarity

Steps:

1. The VP will mark the project as rejected in SharePoint.
2. The VP will mark the project as rejected in Clarity.

13.4.1.2.13 Go To Division's Project Council?

Owners: Management team

Purpose: To decide whether the project charter should be reviewed at the division's project council

Step:

1. They will decide whether this project charter should be reviewed at the division's project council, typically based upon project size (>1000 hours) and other risk criteria, such as the cross-functional nature of the project, impact to the business, and so forth.

13.4.1.2.14 Schedule for Division's Project Council

Owners: Management team

Purpose: To schedule project charters that need to be reviewed in the division project council meetings.

Steps:

1. If it is decided that the project charter will be reviewed at the division's project council, it will be automatically scheduled for division's project council by updating the project council field on the Area Council SharePoint.

2. Division will pull the project charter for the division's project council meetings, based on the Area Council SharePoint site.

13.4.1.2.15 Notify Project Leader of Status and Next Steps (via E-Mail)

Owners: Area Council

Purpose: To notify the project leader of the status of the project and the next steps

Steps:

1. The project leader will receive an e-mail telling him or her whether his or her project was approved or rejected.

2. The e-mail will contain any necessary next steps. For example, if the project is approved, the project leader will be asked to enter his or her review date for requirements into the SharePoint.

13.4.1.3 Failure Mode and Effect Analysis (FMEA)

The team created a failure mode and effect analysis, brainstorming potential failures in the project charter review process. The FMEA is shown in Figure 13.10 with the Pareto chart prioritizing the failure modes by the risk priority number (RPN) shown in Figure 13.11. The highest RPN based on the severity, occurrence, and detection, included resources not being available, a project not getting marked as approved, a scorecard not being created, and a project charter not being reviewed by the team prior to being reviewed by the Area Council team. We identified and incorporated a recommended action into the process and procedures based on the potential failures.

13.4.1.4 Process Analysis

A process value analysis was performed to assess which of the activities provided value to the process. Inherently, the review of the project charter is an inspection step, if the training is done well, the appropriate skills would be transferred to the project charter preparers and a review step would not be necessary. However, some of the value of the review is to communicate which projects are being done across the area, and to be able to allocate resources across the entire area. The activities in the process that were defined as value added are the actual decisions to approve or reject the project, the Area Council review held with the senior vice president, the vice presidents and directors, and the communication to the project

Process Step	Potential Failure Mode	Potential Effects of Failure	Severity	Potential Causes of Failure	Occurence	Current Process Controls	Detection	Risk Priority Number (RPN)	Recommended Action
Review project charter, enter into Area Council (AC) SharePoint, enter scorecard	Submit the project charter without getting it reviewed with their team.	Project charter has errors. Project charter does not explain the problem or identify the scope	5	Lack of training	2	None	9	90	Incorporate director review
	May not identify all errors	Project charter is not high quality	5	Lack of training	10	None	1	50	Scorecard
Fix problem	Preparer may not fix the problem properly	Project charter is not high quality	5	Lack of training	2	None	1	10	Scorecard
Project lead verify review date initiation in Area Council SharePoint by close of business Thursday	Project lead puts in wrong date.	Project charter does not get reviewed	6	Not reading procedures	2	None	5	60	Training
Review project charter and update scorecard and SharePoint.	Project lead does not create the scorecard.	Does not catch errors	10	Not reading procedures	10	None	1	100	Verify before review, in procedure
Enter in scorecard, SharePoint (notify owner)	Project lead does not correct error.	Project charter is not high quality.	10	Lack of engagement	8	None	1	80	Scorecard
Schedule for Area Council (notify project lead to complete project charter action item)	Reviewer misses the project charter and does not get the project on the agenda.	Project can be delayed.	10	Lack of training	1	None	4	40	Training and procedure

Continued

Process Step	Potential Failure Mode	Potential Effects of Failure	S e v e r i t y	Potential Causes of Failure	O c c u r e n c e	Current Process Controls	D e t e c t i o n	Risk Priority Number (RPN)	Recommended Action
Review in Area Council.	Project does not get approved.	Work so far is wasted.	10	Poor scoping, lack of skills, no stakeholder engagement	1	None	1	10	Training
Review in Area Council.	Resources not available	Customer is not satisfied.	10	Lack of visibility or budget	8	None	2	160	Reporting
Mark project as approved in SharePoint and Clarity.	Project does not get marked as approved.	Project is delayed and customer is not satisfied.	10	Mistake	3	None	5	150	Training
Go To project council?	Forget to mark SharePoint for further review.	Cross-divisional dependencies may not be identified.	5	Mistake	2	None	2	20	Verification step

FIGURE 13.10
Failure mode and effect analysis.

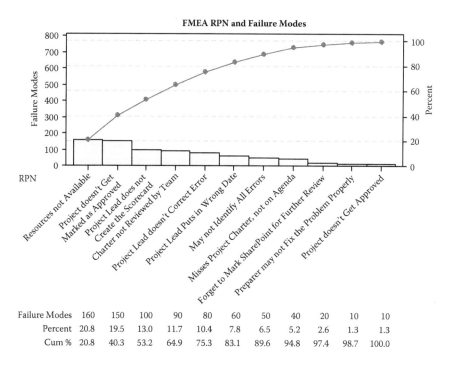

FMEA RPN and Failure Modes

	Failure Modes	Percent	Cum %
Resources not Available	160	20.8	20.8
Project doesn't Get Marked as Approved	150	19.5	40.3
Project Lead does not Create the Scorecard	100	13.0	53.2
Charter not Reviewed by Team	90	11.7	64.9
Project Lead doesn't Correct Error	80	10.4	75.3
Project Lead Puts in Wrong Date	60	7.8	83.1
May not Identify All Errors	50	6.5	89.6
Misses Project Charter, not on Agenda	40	5.2	94.8
Forget to Mark SharePoint for Further Review	20	2.6	97.4
Preparer may not Fix the Problem Properly	10	1.3	98.7
Project doesn't Get Approved	10	1.3	100.0

FIGURE 13.11
Failure mode and effect analysis (FMEA) Pareto chart risk priority number (RPN) priority.

leads of whether the project was approved, deferred, or rejected. The Area Council review provides value from providing communication of work being performed across the area, and the potential to allocate resources across projects, and find and eliminate any project redundancies. The communication of the project approval to the project leads so that the team can move forward on the project also provides value. Only 25% of the activities add value to the process, with 75% of the activities being non-value-added. There is still a great deal of opportunity to incorporate preventive activities and training into the process to further reduce the number of reviews necessary to get a high-quality project charter. The results of the process value analysis combined with the waste analysis results are shown in Figure 13.12.

13.4.1.5 Waste Analysis

A waste analysis was performed on the process. The main types of waste are related to processing embedded in the nature of the process. The project charter review process is being created to provide communication of the work across the entire area and even the entire information systems division, to potentially share resources, and to ensure a high-quality project charter.

Process Step	Value Added	Non-Value Added	Type of Waste
Review project charter, enter into Area Council (AC) SharePoint, enter scorecard		X Inspection	Processing
Fix problem		X Defect	Defect
Project lead verify review date-initiation in Area Council SharePoint by close of business Thursday		X Inspection	Processing
Review project charter and update scorecard and SharePoint.		X Inspection	Processing
Enter deferred in scorecard, SharePoint (notify owner)		X Inspection	Processing
Schedule for Area Council (notify project lead to complete project charter action item)		X Inspection	Processing
Review in Area Council (communicate value of project)	X		Processing
Mark project as approved, deferred, or rejected in SharePoint and Clarity	X		
Schedule for division's project council		X Inspection	Processing
Go to division's project council?		X Inspection	Processing
Schedule for division's project council.		X Inspection	Processing
Notify project lead of status and next steps (via e-mail)	X		

FIGURE 13.12
Process value and waste analysis.

However, there are several levels of review, and the focus of the process should be to incorporate more up-front preventive activities such as training to reduce the number of reviews necessary to get to a high-quality level. When problems are discovered, this is a defect waste. There are many steps in the process that identify defects, and prevention activities should be incorporated to try to reduce or avoid the mistake in the first place. The waste analysis identifying the types of waste for each major step in the process is shown in Figure 13.12.

13.4.1.6 Operational Definitions

The potential metrics were developed to help to ensure the CTS criteria could be met. The voice of process (VOP) matrix (Figure 13.13) summarizes and relates the CTS, process factors that impact the CTS, the operational definition, metrics, and proposed targets. The operational definitions describe how you would specifically measure the metrics that relate to the CTS. To assess a timely process, we will track that the Area Council review is held when scheduled, and the project charters that are scheduled are

Critical to Satisfaction (CTS)	Process Factors	Operational Definition	Metric	Target
Timely process	Procedures followed Management commitment Resources available	Area Council review is held on the scheduled dates, and projects that are scheduled for the agenda are reviewed during the review.	Area Council review is held when scheduled, and projects that are scheduled are reviewed.	100% of projects are reviewed in the identified Area Council review.
High-quality process with metrics	Training Process in place Procedures written, communicated, and followed	Scorecard with content quality criteria and score (see scorecard) Scorecard with format criteria and score (see scorecard)	Content quality percentage Format percentage	Content quality: 80% within 3 months Format: 100% within 3 months
Accurate information	Training Procedures Relationship with business areas	Scorecard with content and format criteria	Content quality percentage Format percentage	Content quality: 80% within 3 months Format: 100% within 3 months
Ability to make decisions, go/no go on projects	Business knowledge of management Quality of project charter	Each project is approved, deferred, or rejected. This would measure the percent approved or rejected, compared to the percent deferred.	Percent of projects approved or rejected the first time (not deferred).	95% (within 3 months of process implementation) of projects that are approved or rejected the first time (0% deferred)
Visibility to program/ project relationships	Program ID is assigned Knowledge of scope of programs and projects	Count of projects that should be related to a program, have the program identified	Count of projects related to programs	80% (within 6 months) of projects that should be related to a program have the program ID.

FIGURE 13.13
Voice of process (VOP) matrix.

reviewed during the session. The target is that 100% of the project charters are reviewed when scheduled.

To assess that a high-quality process with metrics is in place, we developed two initiation scorecards: one to assess the format, and the other to assess the quality of the content. The format initiation scorecard verifies every required field is completed. The content initiation scorecard ensures that the quality of the content in each field meets the standard criteria identified. We used the standard project criteria provided by the division and used them to create the scorecard for each field of the project charter. For the format, each required field was rated as either complete for 1 point, or as a 0 denoting a missing field. There were a total of 30 required fields resulting in a total number of 30 points. The format percentage was calculated as the total number of completed fields divided by the total number of points. For example, if a person missed completing four fields, and completed 26 of the fields, their format percent would be 87% (26/30).

For the content scorecard, a Likert-type rating scale was used, with a scale from 1 (low quality) to 5 (high quality) for each field. Specific semantic definitions were developed for the ratings of 1, 3, and 5. The 2 and 4 ratings are included to allow a rating between the other ratings when the field entry does not quite meet the next higher rating, or the next lowest. There was a total of five points for each field, 1 being the low rating and 5 being the highest rating. There were 12 different fields that were assessed for the content. A perfect content project charter would get a total number of content points of 60, for 100% content. If someone received a three rating on one of the fields, and 5 on all of the others, a total number of points received on the content scorecard would be 58, for a content percentage of 97% (58/60).

The scorecard criteria will be discussed next. The most important fields on the project charter will be discussed along with the criteria for each.

13.4.1.7 Business Opportunity

The business opportunity describes the problem, challenge, or opportunity in the business area that initiated the need for the information systems project. We want to ensure that the business opportunity describes the business problem the project is trying to address.

13.4.1.7.1 Business Opportunity Scorecard Criteria

Format: This is a required field and must be entered.

Content: The following criteria were used to assess the business opportunity.

1. Does *not* explain the business problem
2. Somewhat previous answer, but not quite next answer
3. Explains the business problem/uses abbreviations/grammatical errors
4. Somewhat next answer, but not quite previous answer
5. Explains the business problem/impact to business; one paragraph or less; written in business terms; does not reference a solution; factual representation of what the project is to fix, improve, eliminate, or provide; no abbreviations, grammatical errors

13.4.1.8 Goal

The goal is a statement of how the project will address the identified business problem.

13.4.1.8.1 Goal Scorecard Criteria

Goal: This is a required field and must be entered.

Content: The following criteria were used to assess the goal.

1. Does *not* state how the project addresses the business problem

2. Somewhat previous answer, but not quite next answer

3. Defines how the project addresses the business problem

4. Somewhat next answer, but not quite previous answer

5. Explains the business problem/impact to business; one paragraph or less; written in business terms; does not reference a solution; factual representation of what the project is to fix, improve, eliminate, or provide; no abbreviations, grammatical errors

13.4.1.9 Objective(s)

The objective is a list of high-level bullet points that expand the goal statement and define the boundaries/scope of the project.

13.4.1.9.1 Objectives Scorecard Criteria

Format: This is a required field and must be entered.

Content: The following criteria were used to assess the objectives.

1. Does *not* define the scope of the project; task list

2. Somewhat previous answer, but not quite next answer

3. Defines the scope of the project

4. Somewhat next answer, but not quite previous answer

5. Bullet point list; expands upon the goal statement; defines the boundary/scope of the project; descriptive of future desired state; not a list of tasks

13.4.1.10 Success Criteria

The success criteria identify the end state of the project. The success criteria should be SMART (Specific, Measurable, Attainable, Realistic, and Timely).

13.4.1.10.1 Success Criteria Scorecard Criteria

Format: This is a required field and must be entered.

Content: The following criteria were used to assess the success criteria.

1. Criteria meet 0 of the 5 SMART points.

2. Criteria meet 1 of the 5 SMART points.

3. Criteria meet 2 of the 5 SMART points.

4. Criteria meet 3 of the 5 SMART points.

5. Criteria meet 4 of the 5 SMART points.

6. Criteria meet all 5 SMART points; completes the statement: this project is successful when…; ties back to objectives.

13.4.1.11 Risks

Risks identify factors that can negatively impact the outcome of the project.

13.4.1.11.1 Risks Scorecard Criteria

Format: This is a required field and must be entered.

Content: The following criteria were used to assess the risks.

1. No risk factors identified
2. Somewhat previous answer, but not quite next answer
3. Identifies prioritization, resource, or budget risks only
4. Somewhat next answer, but not quite previous answer
5. Identify factors that can negatively impact the outcome of the project.

13.4.1.12 Assumptions

The assumptions are factors considered to be true without demonstration of proof that could impact the outcome of the project.

13.4.1.12.1 Assumptions Scorecard Criteria

Format: This is a required field and must be entered.

Content: The following criteria were used to assess the assumptions.

1. No assumptions identified
2. Somewhat previous answer, but not quite next answer
3. Identifies prioritization, resource, or budget assumptions only.
4. Somewhat next answer, but not quite previous answer
5. Identifies factors considered to be true (without demonstration of proof) that could impact the outcome of the project.

The concept of the initiation scorecards is to help the project charter authors better understand the criteria for a high-quality project charter, as well as to be used to assess the quality of the charters by the Area Council. Because the measurement against the scorecard criteria can be somewhat subjective, we only use one person to evaluate the quality of the project charters, until we can train others to consistently score the project charters. We would then perform a Gauge R&R (Repeatability and Reproducibility) study to assess the consistency of the measurement system.

13.4.2 Design Phase Presentation

Prepare a presentation (PowerPoint) from the case study exercises that provides a short (10 to 15 minutes) oral presentation of the Design phase deliverables and findings.

13.4.3 Design Phase Case Discussion

13.4.3.1 Design Report

1. Review the Design report and brainstorm some areas for improving the report.
2. How did your team ensure the quality of the written report? How did you assign the work to your team members? Did you face any challenges of team members not completing their assigned tasks in a timely manner, and how did you deal with it?
3. Did your team face difficult challenges in the Design phase? How did your team deal with conflict on your team?
4. Did your instructor and/or Black Belt or Master Black Belt mentor help your team better learn how to apply the Design for Six Sigma tools in the Design phase, and how?
5. Did your Design phase report provide a clear understanding of the root causes of the discipline process? Why or why not?

13.4.3.2 Process Map

1. Was it difficult to create a process map for the process, and also the procedures?

13.4.3.3 Failure Mode and Effect Analysis

1. What other potential failure modes could be identified that were not in the report or in your analysis?
2. How did you determine the recommended actions?

13.4.3.4 Process Analysis

1. Discuss how your team defined whether the activities were value added or non-value added. Was the percent of value-added activities what you would expect for this type of process and why?

13.4.3.5 Waste Analysis

1. What types of waste were prevalent in this process and why?

13.4.3.6 Operational Definitions

1. What other metrics could you identify and measure?
2. Was it difficult to clearly define the operational definition?

13.4.3.7 Design Phase Presentation

1. How did your team decide how many slides/pages to include in your presentation?
2. How did your team decide upon the level of detail to include in your presentation?

13.5 Optimize Phase

13.5.1 Optimize Phase Exercises

13.5.1.1 Optimize Report

Create an Optimize phase report, including your findings, results, and conclusions of the Optimize phase.

1. *Implementation Plan*: Develop an implementation plan for the designed process.
2. *Statistical Process Control*: Develop an example of a control chart that could be used to ensure that the process stays in control.
3. *Process Capability*: Perform a capability analysis to assess whether the process is capable of meeting the target metrics.
4. *Revised Process Map*: Revise your process map to incorporate improvements that will further enhance the process.
5. *Training Plans, Procedures*: Create a training plan, and a detailed procedure for the new process.
6. *Optimize Phase Presentation*: Prepare a presentation (PowerPoint) from the case study exercises that provides a short (10 to 15 minutes) oral presentation of the Optimize phase deliverables and findings.

13.5.2 Optimize Phase

13.5.2.1 Optimize Report

Following is a written report of the Optimize phase for the project charter review process design project, including the key deliverables developed as part of the prior exercises.

The Optimize phase of the IDDOV process is designed to implement the designed process, and then optimize the design by error proofing and further improving the process by seeing what worked and what did not. The main activities of this phase are as follows: implement process, assess process capabilities, and optimize design.

Activity	Responsible	Due Date	Stakeholders Impacted
Develop communication plan for key stakeholders	Process and Metrics team	2/22	All
Distribute new process notice	Process and Metrics team	2/29	All
Hold first Area Council	Process and Metrics team	3/4	All
Assess results, and improvement ideas	Process and Metrics team	3/18	All
Assess process capability	Process and Metrics team	6/17	All
Implement redesigned process	Process and Metrics team	7/17	All

FIGURE 13.14
Implementation plan.

13.5.2.2 Implementation Plan

The team reviewed the process map and procedures with key stakeholders to ensure it met their needs, and aligned with the divisional standards. They then developed an implementation plan, as shown in Figure 13.14. The team also developed a detailed communication plan, shown in Figure 13.15, so they could effectively reach all of the stakeholders so they could understand the new project charter review process. The newly designed process was implemented at the end of February by notifying the entire area through an e-mail with the new process map and detailed procedure. The VPs and directors also communicated the new process to the development teams in their staff meetings and town hall meetings. The first Area Council was held on March 4. The process and metrics team gathered input from the stakeholders, and also held some focus groups to understand any issues and to collect improvement ideas regarding the process.

13.5.2.3 Statistical Process Control

Statistical process control was used to monitor the content and format scorecards by applying a p-chart. The control chart for the first 3 months of format data is shown in Figure 13.16. The format chart for the first 3 months of content data is shown in Figure 13.17. There were many out-of-control points, especially in the first month that the review was running.

13.5.2.4 Process Capability

When we implemented the process, we first baselined the process for the scorecard metrics related to the format and content of the project charter. Figure 13.18 shows the baseline format scorecard percentage of 86%. Figure 13.19 shows the baseline content scorecard percentage of 79%.

The process capability was assessed after 3 months to have enough data available for an adequate sample size. The initiation scorecard metrics were tracked with each area review to assess improvement from a format and

Customers/Stakeholders Communication Plan

	Program	Initiation	Requirements	Technical	Implementation	Postimplementation
Vice president (VP)	• Area program review • Marc's staff meeting • VP staff meetings	• Area Council Document of understanding review • Marc's staff meeting • VP staff meeting	• Area Council DOU review • Marc's staff meeting • VP staff meeting	• Area Council DOU review • Marc's staff meeting • VP staff meeting	• Area Council DOU review • Marc's staff meeting • VP staff meeting	• Area Council DOU review • Marc's staff meeting • VP staff meeting
Directors	• Area program review • Director's staff meetings • Program workshops	• Area Council DOU review • Director's staff meetings	• Area Council DOU review • Director's staff meetings	• Area Council DOU review • Director's staff meetings	• Area Council DOU review • Director's staff meetings	• Area Council DOU review • Director's staff meetings
Managers	• Director's staff meetings • Program workshops	• Director's staff meetings	• Director's staff meetings	• Director's staff meetings	• Director's staff meetings	• Director's staff meetings
Development team:						
Project lead	• Program workshops	• Program/project leader meeting???	• Program/project leader meeting???	• Program/project leader meeting???	• Program/project leader meeting???	• Program/project leader meeting???
Program lead	• Program workshops	• Program/project leader meeting	• Program/project leader meeting	• Program/project leader meeting	• Program/project leader meeting	• Program/project leader meeting
Business analysts (BA)		• BA biweekly meeting	• BA biweekly meeting	• E-mail	• BA biweekly meeting	• BA biweekly meeting
Technical roles		• Project charter workshops	• Gap	• E-mail, need tech meeting?	• E-mail, need tech meeting?	• E-mail, need tech meeting
Division project management office	• Staff meetings • Governance committee • Area Council steering committee	• Staff meetings • Governance committee • Area Council steering committee	• Program Management Office (PMO) staff meetings governance committee • Area Council Steering Committee	• PMO staff meetings • ISD governance committee • Area Council steering committee	• PMO staff meetings • ISD governance committee • Area Council steering committee	• PMO staff meetings • ISD governance committee • Area Council steering committee

Division process engineering	• Staff meetings • Governance committee • Area Council steering committee	• Staff meetings • Governance Committee • Area Council Steering Committee	• Information Systems Development Life Cycle staff meetings?? • ISD governance committee • Area Council steering committee	• ISDLC staff meetings?? • ISD governance committee • Area Council steering committee	• ISDLC staff meetings • ISD governance committee • Area Council steering committee
Other areas in division	• Governance committee • Area Council steering committee	• Governance committee • Area Council steering committee	• BA biweekly meeting? • ISD governance committee • Area Council steering committee	• ISD governance committee • Area Council steering committee	• BA biweekly meeting? • ISD governance committee • Area Council steering committee

FIGURE 13.15
Communication plan..

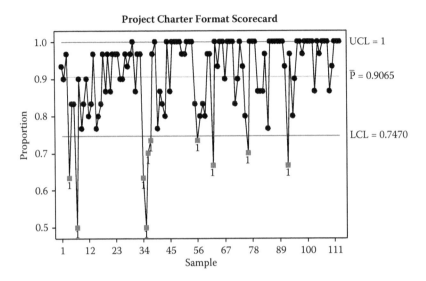

FIGURE 13.16
Format scorecard control chart, with out-of-control points, dates 3/4 to 6/3.

content quality perspective. The format percentage and the content percentage against the scorecard criteria were graphed on p-charts. The quality characteristic used for the p-charts was percent of criteria met for both the format and content scores, and for each project charter reviewed per session. This data were collected for 3 months. There were several points

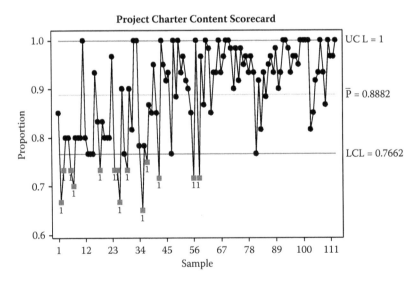

FIGURE 13.17
Content scorecard control chart with out-of-control points, dates 3/4 to 6/3.

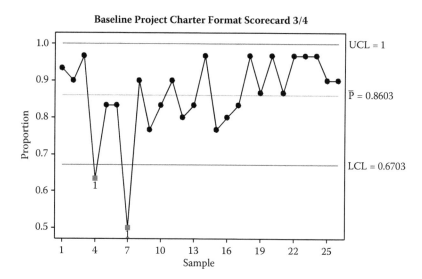

FIGURE 13.18
Baseline format scorecard control chart, date 3/4.

that were out of control during each session, when all of the data were placed on a control chart for the first 3 months' worth of data. Assignable causes were either lack of training, or new project leaders were creating the project charters, so these points were removed to calculate the process capability indices.

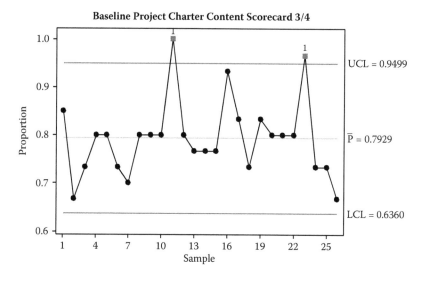

FIGURE 13.19
Baseline content scorecard control chart, date 3/4.

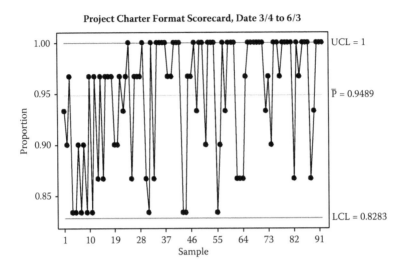

FIGURE 13.20
Format scorecard control chart with assignable causes removed, dates 3/4 to 6/3.

We calculated the *format* process capability to be 95%, after removing the out-of-control points, shown in Figure 13.20.

After the assignable causes were removed, we calculated the *content* process capability to be 96%, as shown in Figure 13.21.

The process capability for a p-chart is the average p value after the process is in control and all of the assignable causes are removed. The process capability

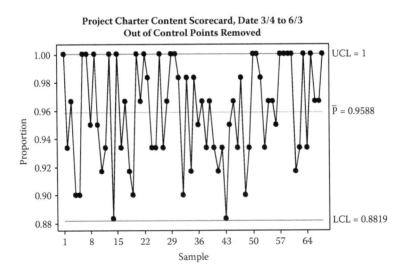

FIGURE 13.21
Content scorecard control chart with assignable causes removed, dates 3/4 to 6/3.

for the project charter *format* is 95%, and the process capability for the project charter *content* is 96%. This equates to a sigma level of about 3.2 to 3.3 sigma, still much room for improvement if Six Sigma is our stretch goal. The 3-month target for the format was 100%, so we were still shy of the target for filling out all of the required fields on the project charter. The 3-month target for the content was 80%, and we have far exceeded the scorecard content target with the process capability of 96%. There is still additional room for improvement related to the project leaders completing all of the required fields.

13.5.2.5 Revised Process Map

We held additional focus groups with the development team stakeholders to understand what worked with the process and what could be improved. We met with the authors of the project charter and project leaders responsible for ensuring that the project charters were reviewed by the Area Council. There were several elements of the process that the focus group attendees liked as follows:

- The visibility and action items provided by the process and the SharePoint site
- The set deadlines and process consistency
- The ability to have input into the process
- Being able to plan the review schedules better
- The scorecard helping you to think through the criteria required on the project charter before sending it on for review (This comment was given by someone who received a perfect project charter score the first time she ever wrote a project charter.)

Some of the improvement ideas from the focus group attendees were as follows:

- Would like to have the scorecard feedback to the authors.
- They are not clear on who is supposed to do what.
- SharePoint navigation is confusing.
- What documents must be attached to the SharePoint?
- Not clear on the review process.
- Challenging to coordinate the functional team reviews with the Area Council review. Timing of the review is difficult (only first and third Tuesdays).

We revised the process to include the following changes:

- In the functional review, we changed the wording from approve to "OK?" to clarify when the project charter is officially approved.

- We combined the format and content scorecard into one document, but kept the ability to report the scores separately. The initial plan was to eventually eliminate the format scorecard when everyone was trained to complete the project charter; however, because the format percentage has not reached the target, we combined the two scorecards for ease of entry but still report on both scores.
- We changed the criteria for the projects that had to also be reviewed in the division's project council to reduce the number of projects that had to be reviewed three different times, down to only twice.
- We moved the due date earlier to better accommodate the volume of project charters that needed to be reviewed.
- We eliminated the need to attach the project charter on the SharePoint site, requiring them to only be uploaded to the project management repository.
- We started to provide the project scorecard feedback directly to the authors. For perfect project charters, we send an e-mail to the project charter author, the project lead, the author's manager, the director, and the VP to share the good news.
- We are tracking the perfect project charters and share those with the management team at the Area Council review.
- We created a project charter workshop and started training project charter authors to further enhance the quality of the project charters.

A revised process map incorporating many of the improvement recommendations is shown in Figure 13.22.

13.5.2.6 Training Plans, Procedures

We developed the project charter workshops, and started training with the business analysts on the development teams, who create a large number of the project charters. The initial pilot workshop went extremely well. We incorporated suggestions for the workshop to improve the workshop material. We revised the procedures with the revised process ideas.

13.5.2.7 Optimize Phase Presentation

The Optimize phase presentation can be developed by the project team.

13.5.3 Optimize Phase Case Discussion

13.5.3.1 Optimize Report

1. Review the Optimize report and brainstorm some areas for improving the report.

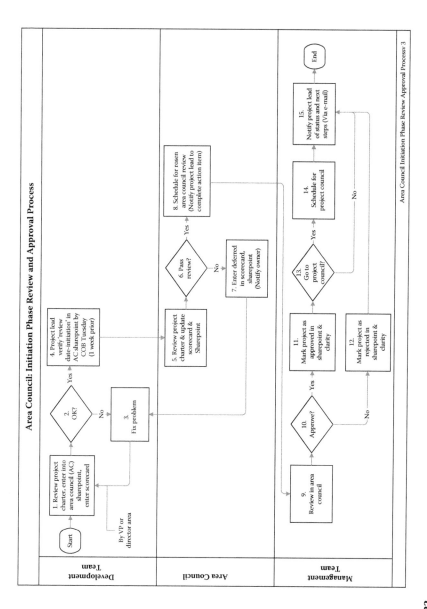

FIGURE 13.22
Revised process map.

2. How did your team ensure the quality of the written report? How did you assign the work to your team members? Did you face any challenges of team members not completing their assigned tasks in a timely manner, and how did you deal with it?

3. Did your team face difficult challenges in the Improve phase? How did your team deal with conflict on your team?

4. Did your instructor and/or Black Belt or Master Black Belt mentor help your team better learn how to apply the Design for Six Sigma tools in the Improve phase, and how?

5. Did your Optimize Phase Report provide a clear understanding of the root causes of the discipline process? Why or why not?

6. Compare your Optimize report to the Optimize report in the book. What are the major differences between your report and the author's report?

7. How would you improve your report?

13.5.3.2 Implementation Plan

1. How must the culture be considered in an implementation plan?

2. How must the communication be considered in an implementation plan?

3. How did your Lean Six Sigma team identify the timings for when to implement your recommendations?

13.5.3.3 Statistical Process Control (SPC)

1. How does SPC help us to control the process?

13.5.3.4 Process Capability

1. Why is it important to assess process capability?

2. Why is it important to ensure that your process is stable before assessing process capability?

13.5.3.5 Revised Process Map

1. Compare your future state process map to the one in the book. How does it differ? Is yours better, worse, the same?

13.5.3.6 Training Plans, Procedures

1. How did you determine which procedures should be developed?

2. How did you decide what type of training should be done?

13.5.3.7 Optimize Phase Presentation

1. How did your team decide how many slides/pages to include in your presentation?
2. How did your team decide upon the level of detail to include in your presentation?

13.6 Validate Phase

13.6.1 Validate Phase Exercises

Validate Report: Create a Validate phase report, including your findings, results, and conclusions of the Validate phase.

Dashboards/Scorecards: Create a dashboard or scorecard for tracking and controlling the process.

Mistake Proofing: Create a mistake-proofing plan to prevent errors from occurring in the process.

Hypothesis Testing/Analysis of Variance (ANOVA): Using the data in the "Project Review Data.xls" spreadsheet, perform the appropriate hypothesis test or ANOVA to compare the scorecard quality between the vice presidents to determine if there is a difference in scorecard quality between the VP areas.

Replication Opportunities: Identify some potential replication opportunities within or outside the division to apply the same or a similar process.

Validate Phase Presentation: Prepare a presentation (PowerPoint) from the case study exercises that provides a short (10 to 15 minutes) oral presentation of the Validate phase deliverables and findings.

13.6.2 Validate Phase Written Report

13.6.2.1 Validate Report

Following is a written report of the Validate phase for the project charter review process design project, including the key deliverables developed as part of the prior exercises.

The purpose of the Validate phase of the IDDOV process is to design, develop, and incorporate controls into the improved processes. The main activities of this phase are to validate process; assess performance, failure modes, and risks; and iterate design and finalize.

> **Format: All fields complete**
> - Baseline: 86.3%; 96.51% (+10.23%)
>
> Content: Meaningful entries in fields
> - Baseline: 79.3%; 93.73% (+14.41%)
>
> Number Total Perfect Project Charters: 18

FIGURE 13.23
Dashboard.

13.6.2.2 Dashboards/Scorecards

The dashboard that is reviewed with management at the start of each Area Council review is shown in Figure 13.23. It shows the initial baseline percent for the format and content scorecard, and the current percentage for the project charters reviewed in the current Area Council cycle, as well as the overall improvement since the baseline. The dashboard also shows the total number of perfect project charters, those that received 100% on both the format and content.

13.6.2.3 Mistake Proofing

To further mistake-proof the process, we developed the following error-proofing ideas:

- Once the project charter author creates the item on the SharePoint, send him or her an e-mail if he or she did not create the scorecard, and encourage the author to revise the project charter based on the scorecard feedback. This can help to improve the project charter before the Area Council review.
- Place a notice on the SharePoint and send an e-mail to notify everyone of the due date so the authors do not submit the project charters late.
- Provide additional project charter workshop training to prevent project charter errors.
- We asked for a program identifier field to be added to the project charter to more easily identify when a project should be associated with a program.
- We added navigational information directions on the SharePoint to reduce confusion identified in the focus group.

13.6.2.4 Hypothesis Testing/Analysis of Variance (ANOVA)

13.6.2.4.1 Hypothesis Testing Between Vice President Areas

After the first 3 months of running the process, we wanted to determine if there was a difference in the project charter format and content scores by the vice presidents' areas, because our VPs are naturally competitive. We first needed to assess whether the format and content scores were normally distributed to determine which statistical test should be used to compare the scores across the VP areas. If the distributions were normal, we could use an ANOVA test; if not we would need to use a nonparametric test such as Kruskal-Wallis or Mood's Median tests.

We performed a normality test in Minitab®, with the null hypothesis being that the data are normal. We received a p-value of .005 for the both the format and content scores. If p is low, the null hypothesis must be rejected. We rejected the null hypothesis and concluded we do not have a normal distribution for the format or the content scores. The histograms for the data are shown in Figures 13.24 and 13.25.

We next tested whether the variances were equal using the Levene's test for the format and scorecard data. The p-value for the format scores was .882, and for the content scores was .724, so we failed to reject the null hypothesis and concluded the variances are equal. We then performed a Mood Median test because it handles outliers better than the Kruskal-Wallis test to test where the median format and content scores are different across the different VP areas. For the format scores, the P-value was .450, so we failed to reject the null hypothesis and concluded the medians were not significantly different. The medians for each of the VP areas were 29 out of 30 on the format scorecards. The Minitab results are shown in Figure 13.26.

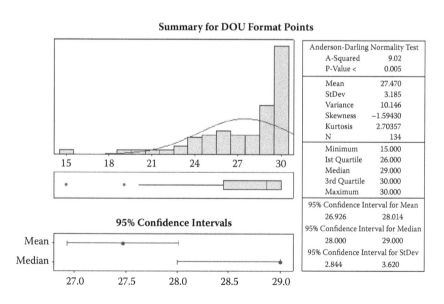

FIGURE 13.24
Histogram of format scores, dates: 3/4 to 5/16.

Summary for DOU Content Points

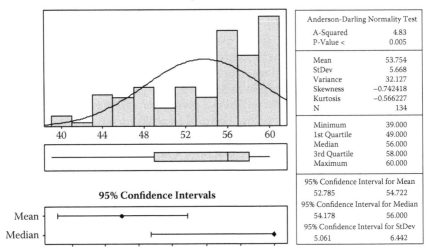

Histogram of content scores, dates: 3/4 to 5/16.

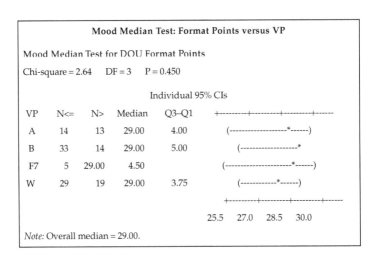

FIGURE 13.26
Format scorecard hypothesis test by vice president (VP) area.

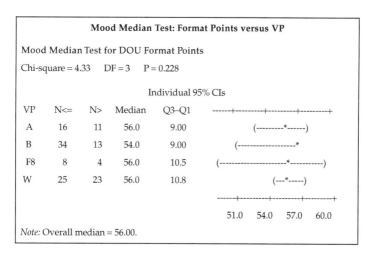

FIGURE 13.27
Content scorecard hypothesis test by vice president (VP) area.

For the content scores, the P-value was .228, so we did not reject the null hypothesis and concluded the content scores are not significantly different across the VP areas. The overall median was 56 out of 60 on the content scorecard. The Minitab results are shown in Figure 13.27.

13.6.2.4.2 Hypothesis Tests from Initial Baseline Results

We wanted to understand if there was an improvement in the format and content scores 3 months after the process was optimized and implemented.

Format Scorecard: We first tested if the variances between the baseline and the more recent scores were equal with a Levene's test. With a null hypothesis that the variances are equal and a p-value of .107, we failed to reject the null hypothesis and concluded the variances are equal. We then performed a Mann-Whitney test to determine if there was a difference between the baseline and later format scores. The null hypothesis was that there is no difference between the baseline and the last 2 months of Area Council results. The conclusion was that the test is significant at 0, so we concluded there is a difference between the baseline format median score (26.5) and the last 2 months of results (30.0), showing a significant improvement in the format scorecard results. The Minitab results are shown in Figure 13.28.

Content Scorecard: We tested if the variances were equal with a Levene's test. With a null hypothesis that the variances were equal and a p-value of .235, we failed to reject the null hypothesis and concluded the variances were equal. We then performed a Mann-Whitney test

---

Mann-Whitney Test and CI: Format 3/4, Format 5/6 to 6/17

	N	Median
DOU Format 3 4	26	26.500
DOU Format Rest	59	30.000

Notes: Point estimate for ETA1-ETA2 is –3.000. 95.1% CI for ETA1-ETA2 is (–4.001, –1.001); W = 631; Test of
ETA1 = ETA2 versus ETA1 not = ETA2 is significant at 0.0000. The test is significant at 0.0000
(adjusted for ties).

FIGURE 13.28
Statistical test for format scorecard, baseline versus last 2 months.

to assess if there was a difference in the content scores between the baseline period and the last 2 months. The null hypothesis was that there was no difference between the baseline and the last 2 months of Area Council results.

The test was significant at 0, so we concluded there was a difference between the baseline median content score (48) and the last 2 months (57), showing a significant improvement in the format scorecard results. The Minitab results are shown in Figure 13.29.

We optimized and validated our new project charter review process.

13.6.2.5 Replication Opportunities

The concept of incorporating the content and format scorecards would be very effective in any similar process, where there is great value in clearly defining and measuring against specific criteria for qualitative information. This encourages assessing knowledge processes, where knowledge is elicited and presented to gain approval to move forward on an information systems project.

This particular Area Council review process and procedures were also adopted in the other senior vice presidents' areas in the division.

Mann-Whitney Test and CI: Content 3/4, Content 5/6 to 6/17

	N	Median
DOU Content 3 4	26	48.000
DOU Content Rest	59	57.000

Notes: Point estimate for ETA1-ETA2 is –10.000. 95.1% CI for ETA1-ETA2 is (–12.000, –8.000); W = 496.5; Test
of ETA1 = ETA2 versus ETA1 not = ETA2 is significant at 0.0000. The test is significant at 0.0000
(adjusted for ties).

FIGURE 13.29
Statistical test for content scorecard, baseline versus last 2 months.

13.6.3 Validate Phase Presentation

Prepare a presentation (PowerPoint) from the case study exercises that provides a short (10 to 15 minutes) oral presentation of the Validate Phase deliverables and findings.

13.6.3.1 *Validate Phase Case Discussion*

1. Validate Report:
   a. Review the Validate report and brainstorm some areas for improving the report.
   b. How did your team ensure the quality of the written report? How did you assign the work to your team members? Did you face any challenges of team members not completing their assigned tasks in a timely manner, and how did you deal with it?
   c. Did your team face difficult challenges in the Validate phase? How did your team deal with conflict on your team?
   d. Did your instructor and/or Black Belt or Master Black Belt mentor help your team better learn how to apply the Design for Six Sigma tools in the Validate phase, and how?
   e. Compare your Validate report to the Validate report in the book, what are the major differences between your report and the author's report?
   f. How would you improve your report?
2. Dashboards/Scorecards:
   a. How would your dashboard differ if it was going to be used to present to just the SVP area or to the entire division?
3. Mistake Proofing:
   a. How well did your team assess the mistake-proofing ideas to prevent errors?
4. Hypothesis Tests, Design of Experiments:
   a. How did you assess the improvement for the CTS?
5. Replication Opportunities:
   a. How did your team identify additional replication opportunities for the process within and outside the information system division?
6. Validate Phase Presentation:
   a. How did your team decide how many slides/pages to include in your presentation?
   b. How did your team decide upon the level of detail to include in your presentation?

14

Information Technology (IT) System Changes—A Design for Six Sigma Case Study

Patricia Long, Jamison Kovach, David Ding,
Elizabeth Cudney, and Sandra Furterer

CONTENTS

14.1 Project Overview

Accounting firms provide financial services, such as bookkeeping, tax returns, and audit services. Unfortunately, employees sometimes encounter problems with information technology (IT) systems that interrupt their work. While the IT department is responsible for fixing these problems, their ability to do so is often hindered by the lack of a formal IT change management system. Such a system would improve the timely identification of prior changes to IT systems that may be related to current problems faced by the IT department, thereby reducing unplanned downtime and IT staff frustration. This work describes a case study that used the Design for Six Sigma (DFSS) methodology to establish a process for effectively managing IT system changes for a mid-size accounting firm. This structured design approach provided an underlying framework for this organization to translate users' needs/expectations into the design of a new system that helped to improve the communication and awareness of system changes within the IT department.

The purpose of this project is to apply DFSS within a small IT department at a mid-size accounting firm. This case specifically demonstrates how the five-phase DFSS methodology, known as IDDOV (Identify-Define-Design-Optimize-Verify), was used as an underlying framework to establish an IT change management process for this department. Change management is the process responsible for controlling and managing requests to change aspects of the IT infrastructure or service and includes activities such as software upgrades, server migrations, data moves and deletions, alterations to file permissions, and the creation and modifications of user accounts. When this project began, the department's approach to managing IT system changes was very informal and ineffective. IT staff simply communicated changes to one another verbally or through e-mail. As a result, changes were often not communicated well or were not acknowledged by others. Hence, the lack of a formally defined IT change management system meant there was no way to determine what system changes had been made that may be responsible for future problems faced by the IT department. The next section describes the company and department in which this case study was conducted as well as the motivation for undertaking the DFSS project. The work completed in each phase of the project is then explained in detail.

The IT department within this organization supports the firm by ensuring that critical software applications and necessary equipment are available and functioning properly so that employees can complete their work tasks. Unfortunately, employees sometimes encounter problems with IT systems that hinder the daily productivity of the organization. When the necessary IT systems are not available, it interrupts the flow of work for employees, thereby potentially causing a loss of revenue. It is the responsibility of the IT department to maintain these systems and fix these problems. This department is small and consists of only four employees—two IT staff, one external consultant, and the IT manager, who reports directly to the managing partner of the firm. At the time of this case study, the IT department identified unplanned IT system downtime as one of their major opportunities for improvement. They felt strongly that their ability to address problems that cause downtime was hindered by the lack of a formal IT change management system within their department.

14.2 Identify Phase

14.2.1 Identify Phase Activities

1. *Develop Project Charter*: Use the information provided in the Project Overview section to develop a project charter for the DFSS project.

2. *Establish Team Ground Rules and Roles*: Establish the project team's ground rules and team members' roles.

3. *Develop Project Plan*: Develop your team's project plan for the DFSS project.

14.2.2 Identify

14.2.2.1 Project Charter

The first step was to develop a project charter.

Project Name: IT Change Management System

Project Overview: Design a process for effectively managing IT system changes, which will result in increased productivity and improved customer satisfaction.

Problem Statement: The IT department does not have a formal process for managing IT system changes, which results in unplanned downtime, security risks, compliance issues, and frustration for the IT staff—all of which negatively impact organizational productivity.

Customer/Stakeholders: All members of the organization

Goal of the Project: Reduce unplanned downtime, improve IT staff morale, improve IT security and compliance issues

Scope Statement: The DFSS team will examine and recommend changes to the major process steps for the IT change management system. The investigation is limited to this process.

Projected Financial Benefit(s): Improved productivity, customer satisfaction, and bottom-line performance

14.2.2.2 *Team Ground Rules and Roles*

The team informally developed several ground rules for the project:

- *Conflict Management*: All team members will be respectful to the other members of the team. All team members will give and receive feedback in a professional manner.
- *Attendance/Tardiness*: Weekly meetings will be held and attendees are expected to be punctual.
- *Participation*: Participation at weekly meetings is expected unless a situation arises where a team member is unable to attend, which may be due to illness or an out-of-town trip.

14.2.2.3 *Project Plan*

A Gantt chart was used to keep the project development cycle on track and ensure that all key steps in the process were accomplished. This chart includes dates and tasks that are essential for appropriate usage of time management during the scheduled period of work.

Also as part of the Identify phase, a project plan was developed. This plan, the execution of which is described in the following sections, included the following steps:

1. Map the current process using a Suppliers-Input-Process-Output-Customers (SIPOC) diagram and a service blueprint.
2. Conduct interviews with all IT staff to determine the needs for the new process.
3. Organize the needs obtained from interviews using an affinity diagram.
4. Prioritize the needs for the new process through a needs assessment survey.
5. Benchmark other organizations to determine initial design ideas.
6. Conduct brainstorming sessions with all IT staff to determine additional design ideas.
7. Organize preliminary design ideas using an affinity diagram.
8. Determine the requirements for the new process using a needs-metrics matrix.
9. Establish baseline measurements for process requirements through a user survey.
10. Select the best design idea using a concept selection matrix.
11. Pilot test the design of the new process.
12. Determine how well the new process fulfills the needs for which it was designed.

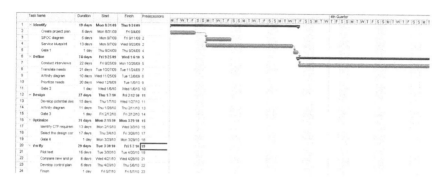

FIGURE 14.1
Project Gantt chart.

To map the operations within the IT department at a high level, the SIPOC diagram shown in Figure 14.2 was created. This diagram shows that the work completed by the IT department, which mainly involves maintaining systems and fixing problems, directly affects company employees, IT staff, and consultants, and indirectly affects the firm's clients. In essence, the IT department provides an internal service within the company to support the firm's overall operations.

To explore the work processes within the IT department in further detail, a service blueprint was created for online change requests, as shown in Figure 14.3. A service blueprint is a process-oriented design technique

Suppliers	Inputs	Process Steps	Outputs	Customers
Help Desk staff	User issues	Receive requests via e-mail/phone	Problem resolved	Company
Information technology (IT) manager	Staff experience	Gather customer information	Standard operating procedure	Employees
Training department	Staff training	Classify issue	Ticket closed and logged documentation	Company clients
External consultants	Staff availability	Analyze issue	Resolution added to FAQ database	IT staff
Vendors	Frequently asked questions (FAQ) database	Solve problem	New configurations/ versions	External consultants
Human resources (HR) department	User availability	Manage requested changes	New hardware/ software installed	
Users	Hardware/software	Install new hardware/software		
Staff	New/terminated users	Perform system maintenance		

FIGURE 14.2
Information technology (IT) department operations.

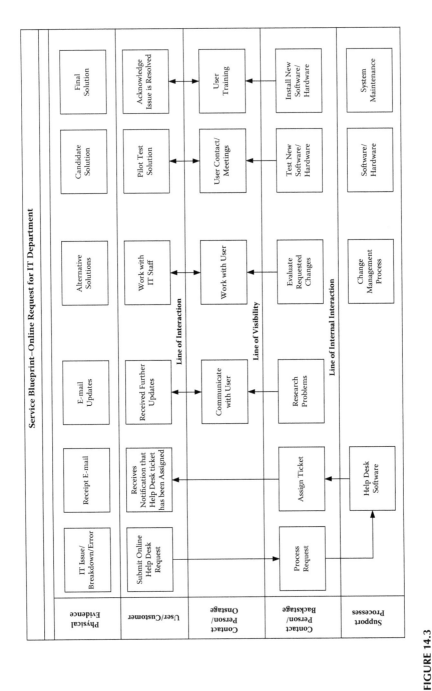

FIGURE 14.3
Flow of work for information technology (IT) help requests submitted online.

intended specifically for documenting and improving service processes that is often used in quality improvement projects. This special type of process map provides a visual representation of the service process that looks somewhat like a swim-lane diagram and denotes the lines of interaction (i.e., where the service provider interacts with the customer), visibility (i.e., where the visibility of the service process by the customer ends), and internal interaction (i.e., where the behind-the-scenes operations within the service process occur). This approach to process mapping specifically differentiates between "onstage" (i.e., aspects of the service process that are visible to the customer) and "backstage" (i.e., aspects of the service process that are not visible to the customer) activities performed by the service provider, which highlights the customer's role in the process (Bitner, Ostrom, and Morgan, 2008). In this case, the onstage activities performed by the IT department include all communications with the users, and the backstage activities include researching solutions to problems, evaluating requested changes, and testing and installing new hardware and software. In addition, the service blueprint shows each point in the process where the employees interact with the customer/user (Bitner, Ostrom, and Morgan, 2008). Here, the IT staff interacts with users when they work to resolve IT issues, which includes developing/pilot testing solutions and training users on new solutions. Mapping these operations helped to organize the collective thinking of the design team so that each member had a detailed understanding of the work processes involved in this project.

14.2.3 Identify Phase Case Discussion

1. DFSS Project Charter: Review the project charter presented.

   a. A problem statement should include a view of what is going on in the business and when it is occurring. The project statement should provide data to quantify the problem. Does the problem statement provide a clear picture of the business problem? Rewrite the problem statement to improve it.

   b. The goal statement should describe the project team's objective and be quantifiable, if possible. Rewrite the goal statement to improve it.

   c. Did your project charter's scope differ from the example provided? How did you assess what was a reasonable scope for your project?

2. Project Plan

   a. Discuss how your team developed their project plan and how they assigned resources to the tasks. How did the team determine estimated durations for the work activities?

14.3 Define Phase

14.3.1 Define Phase Activities

1. *Obtain the Voice of the Customer (VOC)*: Interview employees to identify users' thoughts regarding the current inventory management system (i.e., typical uses, advantages, disadvantages, suggested improvements, etc.).

2. *Identify Users' Needs*: Translate the information gathered through interviews into needs statements (i.e., requirements that the new inventory management system should fulfill).

3. *Prioritize Needs*: Develop a survey and distribute it to users to prioritize the needs that the new system should fulfill.

14.3.2 Define

In the Define phase, a user needs analysis was conducted using the approach described by Ulrich and Eppinger (2004). The user's (i.e., employee's) expectations for an IT change management system were obtained through interviews with IT staff, consultants, and other IT system users.

14.3.2.1 Identifying the Voice of the Customer

As shown in Figure 14.4, six employees participated in interviews in which they were asked to describe their typical uses, likes, dislikes, and suggested improvements regarding the current IT change management process. The responses obtained from these interviews were then translated into "needs" that an IT change management system should fulfill, as depicted by the example interview shown in Figure 14.5. The needs elicited from these interviews were then organized into an affinity diagram, as illustrated in

Information Technology (IT) System Level of Use	Type of User			
	IT Staff	Consultant	Employee	Auditor
Frequently	3			
Occasionally		1		
Rarely			1	1

FIGURE 14.4
Interview participants.

Interviewee: #2		Type of User: IT Staff
Question	**Response**	**Interpreted Need**
1. Typical uses (i.e., what do you need an IT Change Management System (CMS) for)?	1. Any impact to users and other administrators	1. The IT CMS can be used for all changes that impact users or other system administrators.
2. Likes of the current IT CMS (e-mail/verbal)	1. Easy to make changes	1. The IT CMS makes it easy to make changes.
	2. Changes can be made quickly	2. The IT CMS helps to make changes quickly.
3. Dislikes of the current IT CMS (e-mail/verbal)	1. No control over changes that other administrators make	1. The IT CMS limits changes made by other systems administrators.
		2. The IT CMS communicates changes made by other system administrators.
4. Suggested improvements	1. None	—

FIGURE 14.5
Interview data sheet example showing needs translation.

Figure 14.6, by a small group of IT staff who participated in the initial round of interviews.

14.3.2.2 Prioritizing Customers' Needs

To prioritize users' needs, the information summarized in the affinity diagram (Figure 14.6) was used to create a survey, which is displayed in Figure 14.7. In this survey, respondents were asked to rate the importance of the needs that the IT change management system should fulfill using a five-point scale, where a rating of 1 represented "undesirable" and 5 represented "critical." Approximately 14 employees participated in this survey, including the IT manager, all IT staff, their consultant, and several employees from outside the IT department. The results obtained from the survey, shown in Figure 14.8, were analyzed to determine the most important users' needs, thus identifying the characteristics/features that the new process for managing IT system changes should fulfill. Analysis of these data using a chi-square test indicated the importance ratings identified through the survey are different from expected values ($\chi^2 = 149.65$, p-value = <0.0001).

In terms of the critical to satisfaction (CTS) for this project, the design team selected to focus on the 10 needs that received the largest percentage of responses in the "highly desirable" and "critical" categories (i.e., ratings 4 and 5), as shown in Figure 14.8. Figure 14.9 shows the importance weighting assigned to these 10 needs based on median survey responses. Focusing on

Supports Change Tracking
- The CMS provides an effective way to track changes
- The CMS provides an effective way to track who made the changes
- The CMS provides an effective way to track communication with vendors
- The CMS organizes records so they can easily be referenced in the future

Ease & Convenience
- The CMS is fast to use (i.e., changes can be made quickly)
- The CMS is easy to use
- The CMS is easy to access
- The CMS is convenient to use

Audit Log Characteristics
- The CMS provides documentation of changes that are made
- The CMS documents the reason for the change
- The CMS documents who made the change
- The CMS documents when changes were made
- The CMS facilitates the maintenance of documentation
- The CMS facilitates consistency in the way changes are made
- The CMS provides an effective way to log changes
- The CMS ensures that changes are auditable
- The CMS provides the ability to audit the system

Facilitates Communication
- The CMS communications when to make changes
- The CMS communicates who needs to make changes
- The CMS facilitates communication with vendors
- The CMS communicates changes made by others
- The CMS communicates the impact of the change
- The CMS ensures that complicated changes are properly understood
- The CMS controls changes made by other system administrators
- The CMS provides a way to classify changes as temporary or permanent

Common Uses
- The CMS can be used to make program changes/upgrades
- The CMS can be used to make emergency changes
- The CMS can be used for all network changes
- The CMS can be used for all changes that impact users or other system administrators

Facilitates Approval Process
- The CMS facilitates the approval of changes
- The CMS utilizes designated authorities to approve changes

Assessment
- The CMS ensures that program changes are controlled
- The CMS ensures that changes made by one team member do not interfere with the work of other team members
- The CMS differentiates changes based on impact (entire network or group of users vs. individual user)

Supports Planning & Scheduling
- The CMS facilitates planning for future changes
- The CMS facilitates scheduling of changes
- The CMS displays implementation timelines
- The CMS displays updates during planned maintenance windows
- The CMS helps to prevent unplanned down time

Facilitates Teamwork
- The CMS facilitates participation by all team members
- The CMS encourages collaboration among team members
- The CMS facilitates teamwork

Ensures Control
- The CMS supports testing system changes
- The CMS ensures data security

Other
- The CMS ensures data integrity
- The CMS is cost-efficient

FIGURE 14.6
Summary of users' needs obtained from interviews.

Instructions: Please indicate how important the features of an information technology (IT) change management system (CMS) listed below are to you, using the following scale:

1. Feature is undesirable. I would not consider an IT CMS with this feature.
2. Feature is not important, but I would not mind having it.
3. Feature would be nice to have, but is not necessary.
4. Feature is highly desirable, but I would consider an IT CMS without it.
5. Feature is critical. I would not consider an IT CMS without this feature.

Rating   **Feature**

The IT CMS:

_____   1. The CMS is easy to use.

_____   2. The CMS communicates who needs to make the change.

_____   3. The CMS makes it easy to find the information that I need.

_____   4. The CMS makes it easy to document changes.

_____   5. The CMS displays updates during planned maintenance windows.

_____   6. The CMS communicates how the change needs to be made.

_____   7. The CMS provides an effective way to track what was changed.

_____   8. The CMS provides an effective way to track when the change was made.

_____   9. The CMS communicates what change needs to be made.

_____   10. The CMS provides an effective way to log planned changes.

_____   11. The CMS provides an effective way to track communication with vendors.

_____   12. The CMS communicates when to make the change.

_____   13. The CMS communicates why the change needs to be made.

_____   14. The CMS communicates changes made by others.

_____   15. The CMS provides an effective way to track why the change was made.

_____   16. The CMS provides a way to classify changes (i.e., temporary, permanent, etc.).

_____   17. The CMS provides an effective way to track how the change was made.

_____   18. The CMS communicates the impact of the change (i.e., who will be affected).

_____   19. The CMS allows changes to be made quickly.

_____   20. The CMS provides an effective way to track who made the change.

_____   21. The CMS facilitates the scheduling of changes (i.e., displays implementation timelines).

_____   22. The CMS makes it easy to make changes.

_____   23. The CMS utilizes designated authorities to approve changes.

_____   24. The CMS facilitates finding information quickly.

_____   25. The CMS provides an effective way to log changes that have been made.

FIGURE 14.7
User needs survey.

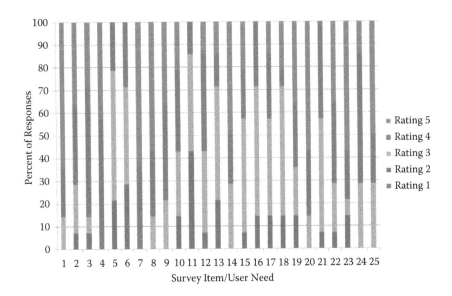

FIGURE 14.8
Results of users' needs survey.

these needs in the subsequent phases of this project helped the design team ensure that the final design fulfills the needs of the work environment for which it was created.

14.3.3 Define Phase Case Discussion

1. How did your team perform the VOC collection? How could VOC collection be improved?

Number	Survey Item Number	Users' Needs (CTQs)	Importance
1	1	The change management system (CMS) is easy to use.	5
2	3	The CMS makes it easy to find the information that I need.	4
3	4	The CMS makes it easy to document changes.	4
4	7	The CMS provides an effective way to track what was changed.	5
5	8	The CMS provides an effective way to track when the change was made.	5
6	14	The CMS communicates changes made by others.	4.5
7	20	The CMS provides an effective way to track who made the change.	5
8	23	The CMS utilizes designated authorities to approve changes.	5
9	24	The CMS facilitates finding information quickly.	4
10	25	The CMS provides an effective way to log changes that have been made.	4.5

FIGURE 14.9
Importance of user's needs based on median importance ratings.

2. What is the value of using an affinity diagram to organize users' needs collected through interviews?

3. Did your team create and distribute a customer survey, and if so, what is the appropriate statistical analysis to perform to identify the importance of the customers' requirements?

14.4 Design Phase

14.4.1 Design Phase Activities

1. *Generate potential design ideas*: Use brainstorming and benchmarking, and then organize design ideas in an affinity diagram.

2. *Develop preliminary design*: Conduct focus groups to discuss the preliminary design ideas listed in the affinity diagram.

14.4.2 Design

In the Design phase, brainstorming and benchmarking were used to generate high-level, potential design ideas. This work involved a total of 11 interviews, as shown in Figure 14.10, to collect ideas from internal as well as external informants in the IT field and beyond. The people interviewed included the IT department manager, internal IT staff, and other firm employees, as well as IT managers from other firms and users of change management systems from other industries, including project managers and firefighters. In these interviews, participants were asked to describe the methods they use to track and communicate changes/updates, how these methods work, and the advantages and disadvantages associated with these methods. The information obtained from these interviews was categorized in an affinity diagram, which is depicted in Figure 14.11. This diagram was then presented to the IT staff as part of a focus group in which this information was discussed, design ideas were further refined, and the team reached agreement about specific design ideas that should be incorporated into the detailed design of the new process for managing IT system changes. These elements are highlighted by bold boxes in Figure 14.11.

Field	Informant		Total
	Internal	External	
Same (information technology, IT)	3	4	7
Different (Non-IT)	2	2	4
Total	5	6	11

FIGURE 14.10
Brainstorming and benchmarking interview participants.

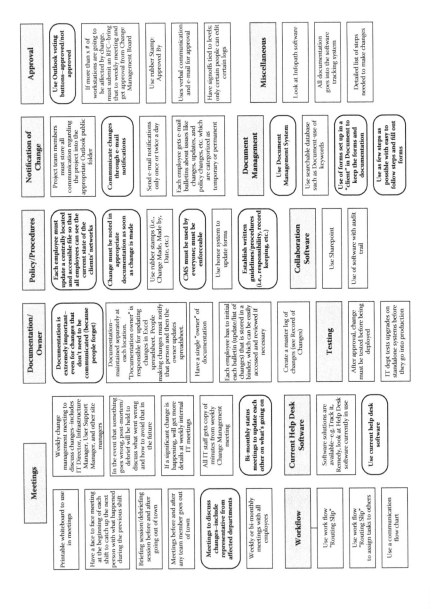

FIGURE 14.11

Summary of preliminary design ideas obtained from benchmarking.

14.4.3 Design Phase Case Discussion

1. How did you generate your design ideas? What other methods besides brainstorming and benchmarking could have been used to generate potential design ideas?
2. What is the value of using an affinity diagram to organize potential design ideas?
3. How did you create your preliminary, high-level design?

14.5 Optimize Phase

14.5.1 Optimize Phase Activities

1. *Identify critical to process (CTP) requirements*: The design team established the technical requirements that the system must fulfill in order to address the users' needs.
2. *Select the design concept*: Various design concepts were assessed against the CTPs using a concept selection matrix to identify the best design idea.

14.5.2 Optimize

The design phase of this project focused on developing a detailed design for the new IT change management system. The design team first developed a general structure, illustrated in Figure 14.12, for the critical elements to be addressed by the new process for managing IT system changes. This framework included developing policies and procedures, addressing communication requirements, addressing documentation requirements, and utilizing software to facilitate operations. The design team felt that the policies and procedures were important because they would dictate the terms of use for the new system. In addition, they believed the other elements included in this framework were essential for guiding the daily operations of the new system.

In this phase, the design team focused on identifying the specific requirements or CTPs that the new process needed to address. These are given in the needs-metrics matrix shown in Figure 14.13. Much like what is done in quality function deployment (QFD) within the House of Quality (Hauser and Clausing, 1988), the dots in Figure 14.13 indicate which needs are addressed by each design requirement. Special attention was given to ensure that each CTS is addressed by at least one CTP. In addition, the importance, units, and baseline measurement for each metric are specified in Figure 14.14. The importance of the CTPs was determined based on the importance of the

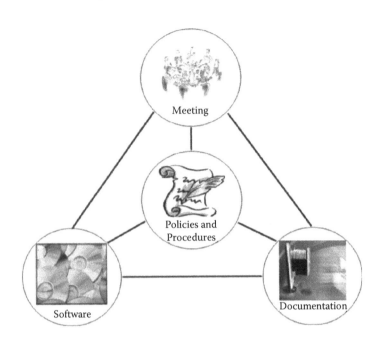

FIGURE 14.12
The structure of the new information technology (IT) change management system.

needs they address. The units selected for each metric reflect the appropriate measurement method for each design requirement. The baseline measurements were also documented to establish a record of the current state of the process before the new system was implemented. Metrics such as "time to document a change" and "time to find information" do not have baseline measurements, because at the time, there was no formal system to document this information within the IT department. Subjective metrics, like "system ease-of-use" and "system effectiveness," were measured through an additional user survey (see Figure 14.15) using a five-point scale, where a rating of 1 represented "falls well short of expectations" and 5 represented "greatly exceeds expectations." This survey was administered to all IT department staff, and the values recorded for the baseline measurement for the remaining metrics were the median rating obtained from the survey and documentation of other aspects of the current system.

Finally, a concept selection matrix was used to select the "best" design concept. As shown in Figure 14.16, the design concepts developed through the analyze phase of this project (i.e., software packages, etc.) are listed across the top of the matrix, and the criteria by which each concept was judged (i.e., design requirements/CTPs) are listed down the leftmost column. One additional selection criterion, "easy to implement," was included to ensure that solutions requiring significant behavioral changes or additional expenses were properly accounted for in this analysis. In this evaluation, each design

	Needs (CTSs)	Metrics (CTPs)	Time to document a change	System ease-of-use	Time to find information	Documents aspects of changes	System effectiveness	Communicates changes	Requires appropriate authority to approve changes
			1	2	3	4	5	6	7
1	The change management system (CMS) is easy to use.		1	1					
2	The CMS makes it easy to find the information that I need.			1	1				
3	The CMS makes it easy to document changes.		1	1		1			
4	The CMS provides an effective way to track what was changed.					1	1		
5	The CMS provides an effective way to track when the change was made.					1	1		
6	The CMS communicates changes made by others.						1	1	
7	The CMS provides an effective way to track who made the change.					1	1		
8	The CMS utilizes designated authorities to approve changes.								1
9	The CMS facilitates finding information quickly.			1	1				
10	The CMS provides an effective way to log changes that have been made.					1	1		

FIGURE 14.13
Needs-metrics matrix.

concept was rated by the design team using a three-point scale, where a rating of 1 represented "will not meet need," 2 represented "will fulfill need," and 3 represented "will exceed expectations for need."

The design concept with the highest score in the selection matrix was the current Help Desk software; hence, the design team decided to use this software package to implement the new IT change management system. Even though the Help Desk software had been used for some time in the IT department, it

Metric Number	Need Numbers	Metric	Importance	Units	Baseline Measurement
1	1, 3	Time to document a change	4.5	Minutes (min.)	—
2	1–3, 9	System ease-of-use	4	5 pt. scale	2
3	2, 9	Time to find information	4	Seconds (s)	—
4	3–5, 7, 10	Documents aspects of changes	5	Binary	No
5	4–7, 10	System effectiveness	5	5 pt. scale	1
6	6	Communicates changes	4.5	Binary	No
7	8	Requires appropriate authority to approve changes	5	Binary	No

FIGURE 14.14
Baseline measurements for design requirements.

> **Instructions:** Please indicate how well the *current* information technology
> (IT) change management system (CMS) meets expectations relative to
> fulfilling the requirements listed below, using the following scale.
>
> The *current* IT change management system:
>
> 1. Falls well short of expectations for this requirement
> 2. Falls short of expectations for this requirement
> 3. Meets expectations for this requirement
> 4. Exceeds expectations for this requirement
> 5. Greatly exceeds expectations for this requirement
>
> **Rating   Requirement**
> _____   1. System ease-of-use (in terms of documenting changes and
>   finding information needed)
> _____   2. System effectiveness (in terms of tracking, documenting, and
>   communicating changes)

FIGURE 14.15
User expectations survey (current system).

is important to note that the change management functions within this software package had not previously been used. As part of the detailed design, the team also decided to continue their biweekly meetings to communicate and discuss changes, maintain network and application documentation in the firm's Document Management software, and establish policies and procedures for the new IT change management process, which included

1. The IT change management system shall be used for all changes that affect production systems.
2. All IT staff will:
   a. Use the change management system
   b. Keep documentation of changes up-to-date
3. Policy compliance audits shall be conducted periodically by the IT manager as well as IT staff (i.e., peer reviews).

Once created, the detailed design was tested through a 2-week pilot study in the IT department. At the conclusion of this pilot test, additional interviews/focus groups were conducted to ensure the design fulfilled the needs of the IT department. Slight adjustments were made based on the results of the pilot study in order to create the final design of the new IT change management system.

14.5.3 Optimize Phase Case Discussion

1. How did you identify the technical requirements and the relationships between customer and technical requirements?

Design Concept

(Scale: 1 = Will Not Meet Need; 2 = Will Fulfill Need; 3 = Will Exceed Expectations for Need)

Selection Criteria	Workflow Software	Sharepoint	Current Help Desk Software	New Help Desk Software	Infopath Software	Outlook	Adobe Forms	Checklists	Rubber Stamps	Meetings
1. Time to document changes	1	1	3	2	1	2	2	2	1	3
2. System ease-of-use (in terms of documenting changes and finding information needed)	3	1	3	2	1	1	3	3	3	3
3. Time to find information	2	2	3	2	1	2	1	1	1	1
4. Aspects of changes can be documented (i.e., what, when, who, etc.)	2	3	3	3	1	1	2	3	1	1
5. System effectiveness (in terms of tracking, documenting, and communicating changes)	2	3	3	3	3	1	1	1	1	1
6. Changes can be communicated	3	2	3	2	3	3	3	2	2	3
7. Approval authority options	2	2	3	3	2	1	2	2	1	3
8. Ease of implementation	2	1	3	1	1	3	2	3	3	3
Total	**17**	**15**	**24**	**18**	**13**	**14**	**16**	**17**	**13**	**18**

FIGURE 14.16
Concept selection matrix.

2. How did you generate your design concepts?

3. How did you determine how your concepts compared using the Pugh Concept Selection matrix?

4. How did you derive the best combination of your design elements from each concept?

14.6 Verify Phase

14.6.1 Verify Phase Activities

1. *Collect data about the new system's performance*: Measure the performance of the new system after implementation.

2. *Compare new system to previous performance*: Use the appropriate statistical analysis methods to compare before and after results to determine whether the new system fulfills the needs for which it was designed.

3. *Develop a control plan*: Establish a way to maintain the improvements achieved through the project.

14.6.2 Verify

Once the final design of the new process for managing IT system changes had been in use for approximately 3 months, several measurements were taken to evaluate whether the new system fulfilled the needs for which it was designed. As shown in Figure 14.17, various aspects built into the new system were documented and compared against baseline measurements to acknowledge the degree of improvement achieved by the new system. Specifically, the "time to document a change" and "time to find information" were determined through several observations of IT department staff using the new system. On average, it now takes 1.65 minutes to document a change and just 9 seconds to find the information they need within the new IT change management system. In terms of "system ease-of-use" and "system effectiveness," again these aspects were measured through a user survey, similar to that shown in Figure 14.15, however it asked IT department staff to indicate how well the "new" system met expectations relative to fulfilling the requirements of ease-of-use and effectiveness—both of which were rated higher for the new system compared to the previous system.

For comparison purposes, survey responses regarding "system ease-of-use" for the previous system versus the new system are shown in Figure 14.18. Chi-square analysis was used to examine proportional changes in ease-of-use ratings between these systems. As shown in Figure 14.19, changes in the IT department staffs' ratings of ease-of-use for the previous system versus

Metric Number	Need Numbers	Metric	Importance	Units	Previous System	New System
1	1, 3	Time to document a change	4.5	Minutes (min.)	—	1.65 min.
2	1–3, 9	System ease-of-use	4	5 pt. scale	2	4.5
3	2, 9	Time to find information	4	Seconds (s)	—	9 s
4	3–5, 7, 10	Documents aspects of changes	5	Binary	No	Yes
5	4–7, 10	System effectiveness	5	5 pt. scale	1	4
6	6	Communicates changes	4.5	Binary	No	Yes
7	8	Requires appropriate authority to approve changes	5	Binary	No	Yes

FIGURE 14.17
Baseline and verification measurements for design requirements.

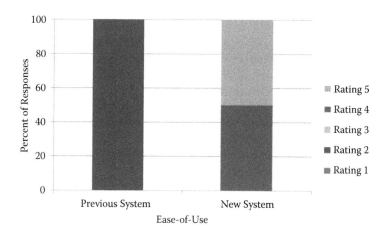

FIGURE 14.18
Survey responses for system ease-of-use.

Ease-of-Use	Counts	Rating Categories* ($n = 4$)					χ^2-value	p-value
		1	2	3	4	5		
Previous system	Count	0	4	0	0	0	8.00	0.0183
	Expected count	0	2	0	1	1		
New system	Count	0	0	0	2	2		
	Expected count	0	2	0	1	1		

FIGURE 14.19
Comparison of system ease-of-use for the new versus the previous system.

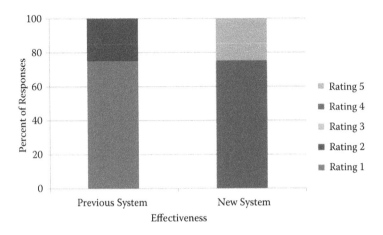

FIGURE 14.20
Survey responses for system effectiveness.

the new system were significant (p-value = 0.0183). Figure 14.20 illustrates survey responses for "system effectiveness" for the two systems. Again, chi-square analysis (see Figure 14.21) indicates that changes in IT department staffs' ratings of system effectiveness were significant (p-value = 0.0460). In other words, there is some evidence to suggest that the new system had a positive impact on improving system ease-of-use and effectiveness.

To maintain the improvements achieved through this project, the design team developed a control plan for the IT department/company that consists of documentation, training, and system monitoring. The instructions for using the new IT change management system were documented in standard operating procedures. The new policies for the change management process described previously were incorporated into the company's Information Technology Infrastructure Protection and Security Policy. All IT department staff, along with other employees involved in the change approval process, were trained on these new policies and procedures. As new employees are hired, this information will be covered as part of their new-hire training program. In addition, the use of the new system will be monitored by the IT

Effectiveness	Counts	Rating Categories* ($n = 4$)					χ^2-value	p-value
		1	2	3	4	5		
Previous system	Count	3	1	0	0	0	8.00	0.0460
	Expected count	1.5	0.5	0	1.5	0.5		
New system	Count	0	0	0	3	0.5		
	Expected count	1.5	0.5	0	1.5	1		

FIGURE 14.21
Comparison of system effectiveness for the new versus the previous system.

manager through regular audits to ensure the system is being used in accordance with the new policies and procedures.

14.6.3 Verify Phase Case Discussion

1. How did your team collect data about the performance of the new system? How could data collection be improved?
2. How did your team identify whether the new system fulfilled the needs of the work environment for which it was designed? What were the appropriate statistical analysis methods to use based on your data?
3. What additional methods could be used to further strengthen your control plan?

14.7 Summary

The work completed in this project illustrates the effective use of the DFSS methodology to create an IT change management system in a mid-sized accounting firm. By using the IDDOV process and associated tools, users' needs were identified and a new process was established that effectively fulfilled the needs of the work environment for which it was designed. The use of this structured design approach achieved several positive outcomes for the organization overall and for the IT department, specifically.

Throughout the DFSS process, all members of the IT department worked together closely to improve the communication of system changes. Because the IT staff are the users of the new system and they helped to design it (i.e., their needs and ideas were considered and integrated into the final design), little to no resistance to using the new system was encountered during implementation. Hence, the new change management system has become an integral part of the daily work processes of the IT department.

Several months after implementation, the new change management system continues to be successfully used by the entire IT department, including their external consultant. The driving force for creating this new system was to stabilize the IT environment for the company. By enhancing communication between all the members of the IT department, providing an approval process for network changes, and maintaining documentation of what changes were made when and by whom, the new change management system has enabled the IT department to reduce overall network downtime and accelerate resolution time of network issues.

The successful implementation of a complete design solution has led to the effective management of changes within the IT department. This new process specifically helped to improve the communication of changes, awareness about changes, and the timely identification of prior changes that may be related to current problems faced by the IT department. As a result, the new system has had a positive impact on reducing unplanned downtime and IT staff frustration, as well as improving productivity and customer satisfaction for the organization as a whole. Based on the success of this project, the IT department is now looking at additional ways to further improve the service they provide to their customers.

References

Bitner, M.J., Ostrom, A., and Morgan, F., Service Blueprinting: A Practical Technique for Service Innovation, *California Management Review*, 50(3): 66–94, 2008.

Hauser, J.R., and Clausing, D., The House of Quality. *Harvard Business Review*, 66(3): 63–73, 1988.

Ulrich, K.T., and Eppinger, S.D., *Product Design and Development*, 3rd ed. New York: McGraw Hill/Irwin, 2004.

15

The Future and Challenges of Design for Six Sigma

Elizabeth Cudney and Sandra Furterer

CONTENTS

This book provided an overview of the Design for Six Sigma methodology, the Identify-Define-Design-Optimize-Validate (IDDOV) methodology, and real-world product and service-oriented case studies applying these methods and tools. This last chapter describes a view into the future with the attempt to project where Design for Six Sigma will evolve over the next decade.

15.1 Applying Lean Methods and Tools to Streamline the Design and Development Life Cycle

Design for Six Sigma (DFSS) is a compilation of several diverse methods and tools that provide a more holistic and integrated toolkit for designing and developing products and services. As new products and services become increasingly complex and multifaceted, it is necessary to more tightly couple and integrate the entire service and product development process with lean design.

A comprehensive methodology is necessary to design a product (application), process, or service right the first time. DFSS is a data-driven quality strategy for designing products and services. Design typically accounts for 70% of the cost of the product, and 80% of quality problems are unwittingly designed into the product (application), process, or service. Therefore, one-third of the budget must be devoted to correcting the problems they plan to create with the first two-thirds of the budget.

The three aspects of improving the product development process include

1. Maximize profitability
2. Minimize time
3. Minimize cost

This must be balanced without compromising value from the customer's perspective.

Lean design/product development aids in identifying and reducing or eliminating waste in the product development process. Lean design focuses on removing waste from all aspects of the product and associated development process before the start of manufacture. Lean design addresses the entire life cycle of a product. More specifically, lean design targets cutting manufacturing costs during the design cycle and accelerating the time-to-market. Kearney (2003) identified the most common forms of waste in product design as shown in Figure 15.1.

Mascitelli (2004) developed five principles of lean design:

Principle 1: Precisely define the customer's *problem* and identify the specific *functions* that must be performed to solve that problem.

Principle 2: Identify the *fastest process* by which the identified functions can be integrated into a high-quality, low-cost product.

Principle 3: Strip away any *unnecessary* or *redundant* cost items to reveal the optimal product solution.

Area of waste reduction	% of design waste
Designs never used, completed, or delivered	Unknown
Downtime while finding information, waiting for test results, etc.	33–50%
Unnecessary documents and prototypes	
Underutilization of design knowledge, for example in costly parts	18%
Over design, such as features customers don't need	8%
Validating manufacturing errors early in the design process	17%
Poor designs producing product defects	15%

(18%, 8%, 17%, 15% bracketed together = 58%)

FIGURE 15.1
Waste in design (Data from UGS PLM Solutions analysis of Tier 1 Automotive suppliers; A.T. Kearney, The Line on Design, 2003. With permission.).

Principle 4: Listen to the voice of the customer *frequently* and *iteratively* throughout the development process.

Principle 5: Embed cost-reduction tools and methods into both your *business practices* and your *culture* to enable cost reduction.

Key areas of waste in product design stem from process delays, design reuse, defects, and process efficiency. Process delays are caused from time lost looking for information, waiting for test results, and waiting for feedback. Waste from design reuse is due to not learning from past design experiences, not reducing unnecessary features, and designs that are never used, completed, or delivered. Defect wastes stem from poor designs and warranty issues. Finally, process efficiency waste is caused from underutilization of design knowledge and not validating manufacturing errors early.

There are several ways to decrease costs in the design cycle by reducing direct material cost, direct labor cost, operational overhead, nonrecurring design cost, and product-specific capital investments. Direct material costs can be reduced by using common parts, design simplification, defect reduction, and parts-count reduction. Direct labor costs can be reduced through design simplification, design for manufacture and assembly, and standardizing processes. Operational overhead can be reduced by increasing the utilization of shared capital equipment and modular design. Nonrecurring design costs can be decreased by standardization, value engineering, and platform design strategies. Product-specific capital investments can be minimized by using value engineering, part standardization, and one-piece flow.

Huthwaite (2004) developed five laws of lean design:

1. Law of Strategic Value: Ensure you are delivering value to all stakeholders during the product's life cycle.
2. Law of Waste Prevention: Prevent waste in all aspects of the product's life.
3. Law of Marketplace Pull: Anticipate change in order to deliver the right products at the right time.
4. Law of Innovation Flow: Create new ideas to delight customers and differentiate your product.
5. Law of Last Feedback: Use predictive feedback to forecast cause-and-effect relationships.

By incorporating lean principles into the Design for Six Sigma methodology, further improvements can be made to the design of a product or service. The product development process can be shortened bringing the product to market faster while still ensuring value to the customer.

15.2 Adapting Enterprise Business Architecture Modeling to Product, Service, and Process Design

With the complexity of products, processes, and technology that impact design decisions, Enterprise Business Architecture modeling can be applied in the future to product and process design as it has been applied to the information system development and design life cycle. Enterprise architecture (EA), also referred to as business architecture (BA) is a relatively recent body of knowledge that comes out of the information systems realm (Bieberstein, Laird, Jones, and Mitra, 2008). Enterprise architecture can provide an enterprise-wide understanding of the business. It attempts to connect the business strategies to planned change initiatives focusing on information technology (IT) projects that can provide tactics to meet the business strategies. In many organizations, the business architecture is documented and developed by the IT organization as a way to understand the business processes. The business processes enable the extraction of key business elements that support required capabilities of the business to meet customers' needs. This provides traceability from the businesses' strategies, to the business requirements, through to the implemented information technology. Demonstrating the alignment between IT initiatives and the business strategies helps to ensure that resources of people, time, and money are applied appropriately.

Business architecture modeling techniques and methods can be used to also provide prioritized alignment with the key strategic initiatives related to design of new products, services, and processes in an enterprise. This can provide alignment between the business strategies and goals and the organization's new product and service development initiatives and goals.

Business architecture helps us to understand the 3- to 5-year strategies of the business. Business architecture provides models that describe the business entities (business processes and relevant business information), their relationships, dynamics, and rules that govern their interaction to achieve enterprise-wide objectives. These same modeling techniques can be used to understand the key strategic initiatives, the products, services, and processes that can be designed to meet the organization's strategic plans, and identify the customer requirements that are met through the product and service designs. The modeling can identify the business capabilities and processes that provide the products and services, linking manufacturing and service processes to the key design decisions.

The elements of the business architecture describe the business enterprise, shown in Figure 15.2 (Furterer, 2011). The business architecture first includes understanding the customers and their needs and expectations. From a product or service design, we can capture the customer needs for the new product or service. Next we capture and document the business strategies and goals,

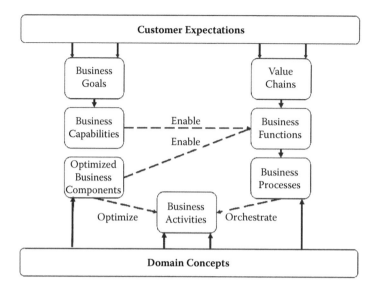

FIGURE 15.2
Business architecture element.

as well as the external and internal influencers on the business. The relationship between the business goals and the business capabilities that support the goals should be understood. We use the value chain showing a chain of activities that provide value to the customer (Porter, 1985) to understand and decompose our processes that help us to meet the new product or service designs. The functional decomposition provides a hierarchical organization of functions and the processes that they include. Each value chain and the subsequent business functions will be used to further decompose the processes. This ensures traceability from the value chains to business processes that provide customer value.

The functional decomposition could be used in product design to decompose and relate the conceptual design customers' requirements to the technical product requirements that meet the customers' needs.

The business capabilities enable the business functions. While the way in which a business implements its processes is likely to change frequently, the basic capabilities of a business tend to remain constant. The business processes and their activities describe the sequence of activities that enable the business to meet the customer's expectations and provide value through the value chains.

The business components are identified to optimize the activities that support the business. These components consist of the activities that require similar people, processes, and technologies. They allow the standardization of the business processes by componentizing the activities that can be used in multiple areas of the business, across many business units and markets (IBM Corporation, 2005).

In product design, the business components could become the product components or even bill of materials that meet the design.

The domain concepts describe the information and roles that exist in the business, that are part of the business processes. The domain conceptual model in product design could describe the product specifications and identify the materials, components, and specifications required in the design of the product.

Incorporating knowledge, methods, and tools from the lean body of knowledge and enterprise business architecture can help us to simplify and deal with the natural growing complexities of product, service, and process design in the future world that we face. You can be part of the journey to enhance product, service, and process design through application of these bodies of knowledge.

References

Bieberstein, Norbert, Laird, Robert G., Jones, Keith, and Mitra, Tilak, *Executing SOA: A Practical Guide for the Service-Oriented Architect*. Upper Saddle River, NJ: IBM/Pearson, 2008.

Furterer, S., *Systems Engineering Focus on Business Architecture: Models, Methods and Applications*, Boca Raton, FL: CRC Press, 2011.

Huthwaite, B., *The Lean Design Solution*, Institute for Lean Design, Mackinac Island, MI, 2004.

IBM Corporation, *Component Business Models, Making Specialization Real*, IBM Business Consulting Services, IBM Institute for Business Value, Somers, NY, 2005.

Kearney, A.T., The Line on Design: How to Reduce Material Cost by Eliminating Design Waste, http://www.atkearney.com.

Mascitelli, Ronald, *The Lean Design Guidebook*, Technology Perspectives, Northridge, CA, 2004.

Porter, Michael E., *Competitive Advantage: Creating and Sustaining Superior Performance*. New York: Free, 1985.

Index

For Product Safety Concerns and Information please contact our EU
representative GPSR@taylorandfrancis.com
Taylor & Francis Verlag GmbH, Kaufingerstraße 24, 80331 München, Germany

www.ingramcontent.com/pod-product-compliance
Ingram Content Group UK Ltd.
Pitfield, Milton Keynes, MK11 3LW, UK
UKHW021624240425
457818UK00018B/719